FOUNDATIONS OF MATHEMATICS

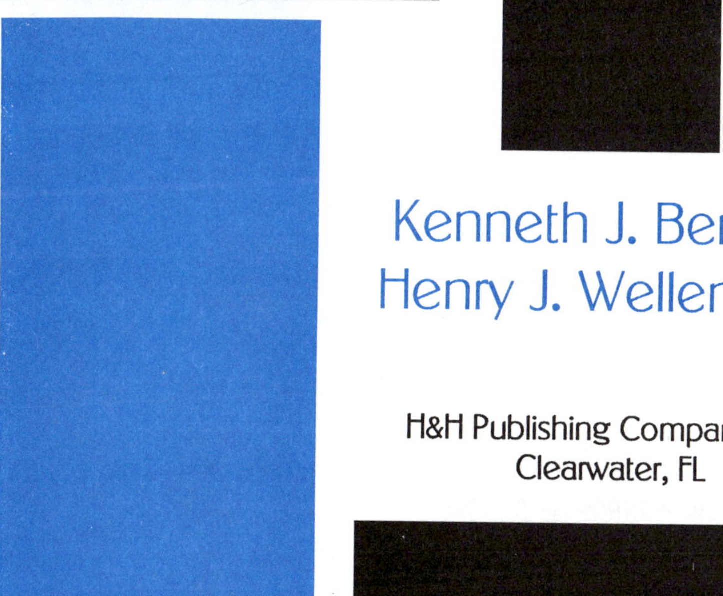

Kenneth J. Bernard
Henry J. Wellenzohn

H&H Publishing Company, Inc.
Clearwater, FL

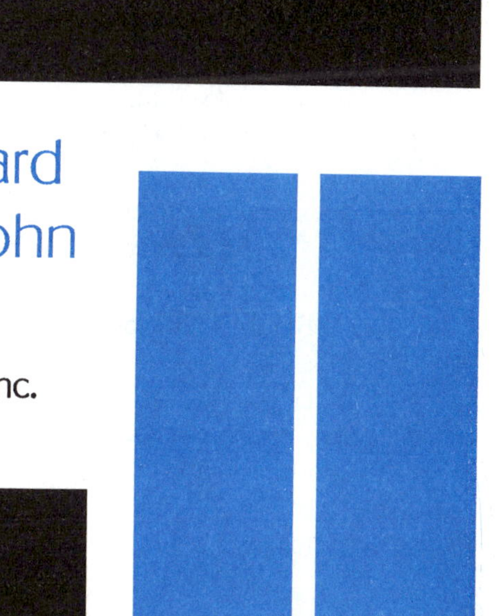

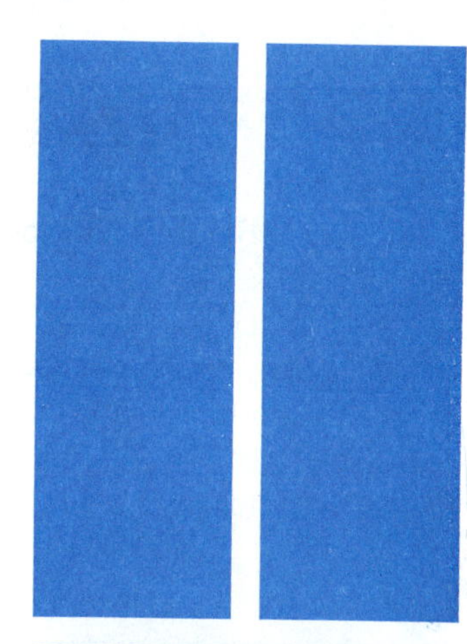

FOUNDATIONS OF MATHEMATICS

by Kenneth J. Bernard & Henry J. Wellenzohn

H&H Publishing Company, Inc.
1231 Kapp Drive
Clearwater, FL 34625
(800) 366-4079
(813) 442-7760
FAX (813) 442-2195
hhcompany@aol.com

Production Supervisor
Robert D. Hackworth

Editor
Karen Hackworth

Editorial Assistant
Priscilla Trimmier

Business Management
Mike Ealy and Sally Marston

ISBN 0-943202-60-4

Printing is the lowest number 10 9 8 7 6 5 4 3 2 1

PREFACE

Foundations of Mathematics is designed to be a self-contained introductory treatment of several topics that are considered to be fundamental to all of mathematics and, consequently, to other disciplines which utilize mathematics. Although the topics considered are not exhaustive, they should be adequate for the intentions and needs of most students. After the study of this book, it is hoped that the student will have established clear, intuitive interpretations as well as precise meanings to the various basic concepts presented. To assist the student along these lines, we have provided examples and counter-examples for the ideas developed in each topic.

This book is intended for use in a course that would be beneficial to students who plan to major in mathematics or other math-related fields; such as, computer science, economics, one of the natural sciences or pre-engineering, and desire a foundation for their subsequent studies. In addition, such a course might also attract students enrolled in liberal arts or business who have an interest in learning about the nature of mathematics. The only prerequisite is that the student have a solid background in high school mathematics. No prior college-level mathematics is required.

Primary emphasis is given to the development of understanding mathematical concepts rather than routine problem solving which one finds in many elementary mathematics books. We have attempted not only to assist the student to think and express himself/herself mathematically but also to foster an appreciation for mathematics. For those students who will pursue more advanced work in mathematics, it is hoped that a level of maturity will be achieved which will enable them to successfully complete the transition from those courses where emphasis is primarily given to problem solving to those courses where emphasis is placed more on abstract, proof-oriented problems.

Chapter 1 is concerned with the study of logic and will show the student how it facilitates communication. Symbolic logic is used to develop a language in which statements, through the use of connectives and quantifiers, provide a concise form of expression. In Chapter 1 we also attempt to convey the spirit and methods of proving the validity of an argument which are so essential to mathematics.

Chapter 2 presents the notion of an object and a set which enter into all branches of mathematics. Properties of sets are demonstrated by applying the algebra of propositions discussed in Chapter 1.

Chapter 3 expands and formalizes the concept of a function with which many students are familiar. Important properties of functions are presented for their own worth as well as aiding the student's understanding of mathematical proofs.

Chapters 4 and 5 are devoted to relations and binary operations. These provide an introduction to the necessary algebraic background for the study of number systems. The correspondence between equivalence relations and partitions of a set are presented.

Chapter 6 deals with some of the more basic concepts in number theory. This not only provides pertinent information for a background in the properties of the integers, but also presents numerous instances in which various methods of proof may be illustrated.

Chapter 7 formalizes the notions of finite and infinite sets. Properties and examples of finite, infinite, denumerable, countable and uncountable sets are provided.

Chapter 8 reviews a few of the basic properties of the real numbers and discusses other facts that should be useful in subsequent courses in mathematics.

At the end of each section there is a problem set. Each problem set consists of several problems, which when solved, should yield additional insights into material discussed. It is hoped that the problems will not only give practice in the use of definitions and theorems, but also a working knowledge of the material.

Solutions to all odd numbered exercises appear at the end of the text. An Instructor's Manual is available which contains the solutions to all the even numbered exercises and chapter examinations.

ACKNOWLEDGMENTS

We wish to thank the editors and staff of H&H Publishing Company, especially Bob Hackworth, Karen Hackworth, and Priscilla Trimmier, for their patience, encouragement, guidance and insightfulness in the preparation of the text.

We are very grateful to the many individuals, especially our students and mathematics colleagues at Niagara University, for the valuable help and constructive criticism rendered us in the development and writing of this text. In particular, we wish to extend our sincere appreciation to Barbara Skye and Sandy Bernard for their patience and expertise in typing the original manuscript and subsequent drafts that evolved into the text. Finally, we wish to thank John Bristol, a former student, who helped with the text layout and whose useful comments and suggestions have improved the final version.

TABLE OF CONTENTS

Chapter 1

LOGIC AND MATHEMATICAL REASONING

Most of the mathematics you have encountered has been mechanical in nature in that the approach to learning a particular concept usually involved repeated applications of that concept to similar problems. In the learning of more advanced mathematics it is essential to be able to derive new knowledge from known facts using careful reasoning. To this extent logic is employed as a useful tool. It plays an important role in the various processes of reasoning which are used in establishing the validity of mathematical statements referred to as "Theorems."

In this chapter we lay the groundwork for those concepts which are to be used in subsequent chapters of this book.

Section 1.1
PROPOSITIONS AND CONNECTIVES

Statements we make, hear or read in everyday life are usually declarative sentences. These sentences may be true, false and, in some instances, both true and false dependant upon an individual's opinion or perspective. We shall restrict our view of statements in mathematics to propositions that we define next.

Definition 1.1.1

A ***proposition*** is a statement to which can be assigned one and only one of the values: "true" or "false."

Using this definition, the statements "Fish are swimmers" and "Seven is an even integer" are propositions since each can be assigned a definite truth value. However, the statement "x is an even integer" is *not* a proposition since no definite truth value can be assigned because the value of x is unknown.

Simple propositions will be denoted by lower case letters of the alphabet; such as p, q, r, etc.

If p is a proposition, then ~p (read "not p") is called the *negation of p*. It is the logical opposite of the proposition p and hence its truth values are reversed.

Definition 1.1.2

If p is a proposition, then the statement "***not* p**," denoted by the symbol **~p**, is a proposition whose truth values are given by the table:

p	**~p**
T	F
F	T

The proposition ~ p is formed by inserting the word "not" at an appropriate place in the proposition p or by preceding the proposition p with the phrase, "It is not true that . . ."

Example 1.1.1

Let p, q and r be defined as follows:

p: Fish are swimmers.
q: Seven is an even integer.
r: Eight is greater than three.

Then the negations of these propositions are:

~p: "It is not true that fish are swimmers," or, "Fish are not swimmers."

~q: "It is not true that seven is an even integer," or, "Seven is not an even integer."

~r: "It is not true that eight is greater than three," or, "Eight is not greater than three."

In daily usage there exists a variety of ways to combine statements to form more complex statements. In logic two or more propositions may be combined through the use of what are called *connectives* to form a new statement called a *compound proposition.* The four basic connectives used in logic are: *conjunction* (p and q), *disjunction* (p or q), *conditional* (if p, then q) and *biconditional* (p if and only if q).

The truth value of a compound proposition depends on the truth value of each of its components and on the basic connective(s) being used. Consequently, we can develop a system for analyzing compound propositions. It should be noted that when two simple propositions are combined by a connective, there are 2 x 2 or 4 logical possibilities. In general, when n simple propositions are combined by connectives, there are 2 x 2 x . . . x 2 (n times) or 2^n logical possibilities.

We shall consider each of the four connectives separately. The use of the connective "and" is very similar to everyday usage.

Definition 1.1.3

If p and q are propositions, the statement "**p and q,**" denoted by the symbol **p ∧ q**, is a proposition whose truth values are given by the table:

p	q	p ∧ q
T	T	T
T	F	F
F	T	F
F	F	F

The proposition p ∧ q is called the ***conjunction*** of p and q and is true only if the truth value of each component is true.

Example 1.1.2

Let p, q, r and s be defined as follows:

p: Two is an even integer.
q: Seven is an even integer.
r: Six is divisible by three.
s: Seven is greater than nine.

We shall discuss the truth value of each of the following compound statements.

p ∧ r: "Two is an even integer and six is divisible by three."

Since each of the components p, r has a truth value of true, the proposition p ∧ r is true.

p ∧ q: "Two is an even integer and seven is an even integer."

Since one of the components; i.e., q, has a truth value of false, the proposition p ∧ q is false.

q ∧ s: "Seven is an even integer and seven is greater than nine."

Since each of the components q, s has a truth value of false, the proposition q ∧ s is false.

The connective "or" in everyday usage may be either inclusive (one or the other or both) or exclusive (one but not both). In logic, "or" is only used inclusively.

Definition 1.1.4

If p and q are propositions, then the statement "**p *or* q**", denoted by the symbol **p ∨ q**, is a proposition whose truth values are given by the table:

p	q	p ∨ q
T	T	T
T	F	T
F	T	T
F	F	F

The proposition p ∨ q is called the ***disjunction*** of p and q and is true if the truth value of at least one of its components is true.

Example 1.1.3

Let p, q, r and s be defined as follows:

p: Two is an even integer.
q: Seven is an even integer.
r: Six is divisible by three.
s: Seven is greater than nine.

We shall discuss the truth value of the following compound statements.

p ∨ r: "Two is an even integer or six is divisible by three."

Since each of the components p, r has a truth value of true, the proposition p ∨ r is true.

p ∨ q: "Two is an even integer or seven is an even integer."

Since one of the components; i.e., p, has a truth value of true, the proposition p ∨ q is true.

q ∨ s: "Seven is an even integer or seven is greater than nine."

Since neither of the components q, s has a truth value of true, the proposition q ∨ s is false.

Many mathematical statements utilize the proposition "if p, then q." It is used to mean that whenever the statement p is true, then q must also be true. For example: "If an even integer is divisible by 2, then 8 is divisible by 2." However, in logic the usage of "if p, then q" is somewhat different in that we must assign a truth value to each of the proposition's components. That is, we must consider all logical possibilities. Hence, the cases for which either p or q is false must also be taken into consideration. Consider the example of one of the cases: "If pigs can fly, then pigs have wings." Though "pigs can fly" and "pigs have wings" are both false propositions, the truth of the compound proposition "If pigs can fly, then pigs have wings" is based on whether it is a *reasonable* statement rather than an absolute sense of truth.

Definition 1.1.5

If p and q are propositions, then the statement "***if*** **p,** ***then*** **q,**" denoted by the symbol **p ⇒ q** is a proposition whose truth values are given by the table:

p	q	p ⇒ q
T	T	T
T	F	F
F	T	T
F	F	T

The proposition p ⇒ q is called a ***conditional*** proposition involving p and q and is true unless p (the hypothesis) is true and q (the conclusion) is false.

There are other ways in which the proposition "**if p, then q**" (**p ⇒ q**) can be stated. Some of them are:

"**p** implies **q**"
"**q** if **p**"
"**q** whenever **p**"
"**p** is a sufficient condition for **q**"

Example 1.1.4

Let p and q be defined as follows:

p: It is raining.
q: We shall not play golf.

We shall discuss the truth values of the conditional proposition $p \Rightarrow q$, "If it is raining, then we shall not play golf," by considering the various truth values of the components p, q.

If **p and q are both true**; i.e., "It is raining" and "We do not play golf," then the proposition $p \Rightarrow q$ is a true statement.

If **p is true and q is false**; i.e., "It is raining" and "We play golf," then the proposition $p \Rightarrow q$ is a false statement.

If **p is false and q is true**; i.e., "It is not raining" and "We do not play golf," the proposition $p \Rightarrow q$ is true since there is no indication what we shall do if it is not raining.

If **p and q are both false**; i.e., "It is not raining" and "We do play golf," then the proposition $p \Rightarrow q$ is true again since there is no indication what we shall do if it is not raining.

The proposition $\mathbf{q \Rightarrow p}$ is called the *converse* of the proposition $\mathbf{p \Rightarrow q}$. The table below (columns 3, 4) shows that these two propositions do not have the same truth tables, i. e., the same truth value in each row of the table. In fact, it should be noted from the table that the converse of a true proposition may not be true.

If we regard the statement "p if and only if q" as having the same meaning as "if p, then q" and "if q then p," then we have the following truth table for the proposition $(p \Rightarrow q) \wedge (q \Rightarrow p)$:

p	q	$p \Rightarrow q$	$q \Rightarrow p$	$(p \Rightarrow q) \wedge (q \Rightarrow p)$
T	T	T	T	T
T	F	F	T	F
F	T	T	F	F
F	F	T	T	T

We note that $(p \Rightarrow q) \wedge (q \Rightarrow p)$ is true only when p and q have the same truth value. Hence, the truth value of the compound proposition "p if and only if q" can be summarized in terms of its components p and q in the following definition.

Definition 1.1.6

If p and q are propositions, then the statement "**p *if and only if* q**," denoted by the symbol **p ⇔ q**, is a proposition whose truth values are given by the following table:

p	q	p ⇔ q
T	T	T
T	F	F
F	T	F
F	F	T

The proposition p ⇔ q is called a ***biconditional*** proposition involving p and q and is true only when each component has the same truth value.

The statement "**p if and only if q**" (**p** ⇔ **q**) may be interpreted as stating that p is not only a sufficient condition for q but also a necessary condition for q.

More than one connective can be used in one statement to produce a more complex compound proposition. Using experience with parentheses and other grouping symbols from algebra, these propositions become manageable.

Example 1.1.5

Using Definitions 1.1.2-1.1.6, construct a truth table for each of the following compound propositions.

a. ~(p ∧ ~q)

p	q	~q	p ∧ ~q	~(p ∧ ~q)
T	T	F	F	T
T	F	T	T	F
F	T	F	F	T
F	F	T	F	T

b. (p ⇒ q) ∧ (q ⇒ r)

p	q	r	p ⇒ q	q ⇒ r	(p ⇒ q) ∧ (q ⇒ r)
T	T	T	T	T	T
T	T	F	T	F	F
T	F	T	F	T	F
T	F	F	F	T	F
F	T	T	T	T	T
F	T	F	T	F	F
F	F	T	T	T	T
F	F	F	T	T	T

Exercises 1.1

For 1-8 construct a truth table for each of the following propositions:

1. $p \Rightarrow (\sim p)$

2. $p \vee (\sim p)$

3. $p \wedge (\sim p)$

4. $(p \vee q) \vee \sim p$

5. $\sim (p \vee \sim p)$

6. $p \vee (q \wedge r)$

7. $p \Rightarrow (p \vee q)$

8. $(p \Rightarrow q) \Rightarrow r$

9. Let p, q and r be defined as follows:
   p: John studied diligently.
   q: John passed all his courses.
   r: John graduated with honors.

   Express each of the following English sentences in symbolic form.

   a. If John studied diligently, then John passed all his courses.
   b. John passed all his courses or John graduated with honors.
   c. John graduated with honors if and only if John studied diligently.
   d. John passed all his courses and John did not graduate with honors.

10. Let p, q and r be defined as follows:
   p: The figure is a triangle.
   q: The figure has less than four sides.
   r: The figure contains a right angle.

   Translate each of the following symbolic forms into an English sentence.
   a. $p \Rightarrow q$
   b. $r \wedge q$
   c. $p \vee r$
   d. $(\sim r) \Rightarrow (\sim p)$

Section 1.2
TAUTOLOGIES AND LOGICALLY EQUIVALENT STATEMENTS

In mathematics, some statements are always true, e.g., “If x is a real number, then $x < x + 1$.” Others are always false, e.g., “If x is a real number, then $x^2 < 0$.” Compound propositions may also have this property.

Definition 1.2.1

A compound proposition is called a ***tautology*** if it always has the truth value “true” regardless of the truth values of its components.

Example 1.2.1

Show that each of the following compound propositions is a tautology:

a. $p \Rightarrow (p \vee q)$

p	q	$p \vee q$	$p \Rightarrow (p \vee q)$
T	T	T	T
T	F	T	T
F	T	T	T
F	F	F	T

b. $[(p \Rightarrow q) \wedge \sim q] \Rightarrow \sim p$

p	q	$p \Rightarrow q$	$\sim q$	$(p \Rightarrow q) \wedge \sim q$	$\sim p$	$[(p \Rightarrow q) \wedge \sim q] \Rightarrow \sim p$
T	T	T	F	F	F	T
T	F	F	T	F	F	T
T	T	T	F	F	T	T
F	F	T	T	T	T	T

Example 1.2.1a is one form of what is called the *Additive Law*, while Example 1.2.1b is one form of what is called the *Law of Detachment*. In Section 1.4, we shall discuss the sequencing of propositions and the development of what is referred to as a proof of a valid argument.

The following table lists a few of the more important compound propositions which are tautologies and which will be used in subsequent sections.

Table 1.2.1

Tautology	Name
$p \Rightarrow p \vee q$ $q \Rightarrow p \vee q$	Addition Laws
$p \wedge q \Rightarrow p$ $p \wedge q \Rightarrow q$	Simplification Laws
$[(p \Rightarrow q) \wedge (q \Rightarrow r)] \Rightarrow (p \Rightarrow r)$	Transitive Law
$[(p \Rightarrow q) \wedge p] \Rightarrow q$ $[(p \Rightarrow q) \wedge \sim q] \Rightarrow \sim p$ $[(p \vee q) \wedge \sim p] \Rightarrow q$	Detachment Laws

In the arithmetic of real numbers, various algebraic expressions can be shown to be equivalent; such as, a(b + c) and ab + ac or a(-b) and -(ab). As a consequence, these equivalent expressions may be used as replacements for one another. A similar situation exists with propositions.

Definition 1.2.2

Two propositions p and q are said to be ***logically equivalent*** and we write $\mathbf{p \equiv q}$ if the biconditional proposition $p \Leftrightarrow q$ is a tautology.

Recall that the biconditional proposition $p \Leftrightarrow q$ is true only when each component, p and q, has the same truth value. Hence, $p \equiv q$ if and only if p and q have the same truth table. From the discussion of the biconditional proposition, this will occur when the propositions have the same truth value in each row of their truth table.

Example 1.2.2

a. The following pairs of propositions are logically equivalent and hence may be used as replacements for one another.

i. $p \Rightarrow q$; $\sim q \Rightarrow \sim p$

p	q	$p \Rightarrow q$	$\sim q$	$\sim p$	$\sim q \Rightarrow \sim p$	$(p \Rightarrow q) \Leftrightarrow (\sim q \Rightarrow \sim p)$
T	T	T	F	F	T	T
T	F	F	T	F	F	T
F	T	T	F	T	T	T
F	F	T	T	T	T	T

The proposition $\sim q \Rightarrow \sim p$ is referred to as the *contrapositive* of the proposition $p \Rightarrow q$.

ii. $p \Rightarrow q$; $\sim p \vee q$

p	q	$p \Rightarrow q$	$\sim p$	$\sim p \vee q$	$(p \Rightarrow q) \Leftrightarrow (\sim p \vee q)$
T	T	T	F	T	T
T	F	F	F	F	T
F	T	T	T	T	T
F	F	T	T	T	T

Note that in each of these examples, the same truth value appears in each row of the two statements being compared.

b. The two propositions, $p \Rightarrow q$ and $\sim p \Rightarrow \sim q$, are *not* logically equivalent:

p	q	$p \Rightarrow q$	$\sim p$	$\sim q$	$\sim p \Rightarrow \sim q$	$(p \Rightarrow q) \Leftrightarrow (\sim p \Rightarrow \sim q)$
T	T	T	F	F	T	T
T	F	F	F	T	T	F
F	T	T	T	F	F	F
F	F	T	T	T	T	T

Note that in this example the truth values are not the same in each row of the statements $p \Rightarrow q$ and $\sim p \Rightarrow \sim q$. The proposition $\sim p \Rightarrow \sim q$ is referred to as the *inverse* of the proposition $p \Rightarrow q$.

In Section 1.1, we showed that the proposition $p \Rightarrow q$ and the converse, $q \Rightarrow p$, do not have the same truth tables and hence are not logically equivalent. In Example 1.2.2, we discussed the inverse, $\sim p \Rightarrow \sim q$ and the contrapositive, $\sim q \Rightarrow \sim p$, of the proposition $p \Rightarrow q$. Let us consider the following concrete example.

Example 1.2.3

Let p and q be defined as follows:

p: It is raining.

q: We shall not play golf.

$p \Rightarrow q$: If it is raining, then we shall not play golf.

$q \Rightarrow p$ is the *converse*:

If we do not play golf, then it is raining.

$\sim p \Rightarrow \sim q$ is the *inverse*:

If it is not raining, then we shall play golf.

$\sim q \Rightarrow \sim p$ is the *contrapositive*:

If we play golf, then it is not raining.

The only statement logically equivalent to $p \Rightarrow q$ is the contrapositive $\sim q \Rightarrow \sim p$.

The following table lists a few of the more important propositions which are logically equivalent and which will be used in subsequent sections.

Table 1.2.2

Equivalence	Law
$\sim(\sim p) \equiv p$	Double Negation
$p \vee p \equiv p$ $p \wedge p \equiv p$	Idempotent Laws
$p \wedge q \equiv q \wedge p$ $p \vee q \equiv q \vee p$	Commutative Laws
$(p \wedge q) \wedge r \equiv p \wedge (q \wedge r)$ $(p \vee q) \vee r \equiv p \vee (q \vee r)$	Associative Laws
$p \wedge (q \vee r) \equiv (p \wedge q) \vee (p \wedge r)$ $p \vee (q \wedge r) \equiv (p \vee q) \wedge (p \vee r)$	Distributive Laws
$\sim(p \wedge q) \equiv (\sim p) \vee (\sim q)$ $\sim(p \vee q) \equiv (\sim p) \wedge (\sim q)$	DeMorgan's Laws
$(p \Rightarrow q) \equiv (\sim q \Rightarrow \sim p)$	Contrapositive Law
$(p \Rightarrow q) \equiv \sim p \vee q$	Alternative Definition of Conditional Proposition

Just as some propositions can always have the truth value "true," as in a tautology, we may have propositions that always have the truth value "false."

Definition 1.2.3

A compound proposition is called a ***contradiction*** if it always has the truth value "false" regardless of the truth values of its components.

Example 1.2.4

The proposition $p \wedge \sim p$ is a contradiction.

p	$\sim p$	$p \wedge \sim p$
T	F	F
F	T	F

Exercises 1.2

For 1-3, show each of the compound propositions is a tautology:

1. $(p \wedge q) \Rightarrow p$
2. $[(p \Rightarrow q) \wedge (q \Rightarrow r)] \Rightarrow (p \Rightarrow r)$
3. $p \Rightarrow [q \Rightarrow (p \wedge q)]$

For 4-9, show that each pair of compound propositions is logically equivalent:

4. $(\sim p \vee q)$; $\sim(p \wedge \sim q)$
5. $(p \wedge q) \Rightarrow r$; $p \Rightarrow (q \Rightarrow r)$
6. $p \wedge (q \vee r)$; $(p \wedge q) \vee (p \wedge r)$
7. $p \Rightarrow (q \vee r)$; $[(p \wedge \sim q) \Rightarrow r] \vee [(p \wedge \sim r) \Rightarrow q]$
8. $(p \wedge q) \Rightarrow r$; $[(p \Rightarrow r) \vee (q \Rightarrow r)]$
9. $p \Rightarrow (q \wedge r)$; $[(p \Rightarrow q) \wedge (p \Rightarrow r)]$

10. Given the proposition $(p \wedge \sim q) \Rightarrow r$, find the converse, the inverse and the contrapositive of the given proposition. (Use rules for logically equivalent statements to put your result in simplified form.)

11. Let p and q be defined as follows:
    p: John studied diligently.
    q: John passed all his courses.

    The proposition $p \Rightarrow q$ is: If John studied diligently, then John passed all his courses.

    Express the converse, the inverse and the contrapositive of the proposition $p \Rightarrow q$ in symbolic form and then translate each into an English sentence.

Section 1.3
QUANTIFIERS

As previously mentioned, the statement "x is an even integer" is not a proposition since no definite truth value can be assigned because the integer value of x is not specified. When a specific integer value is used to replace x, the statement can then be assigned a truth value and thus becomes a proposition. Such statements as "x is an even integer" are referred to as formulas.

Definition 1.3.1

Let X be a collection of objects. A ***formula*** defined on X and denoted by **p(x)** is a statement such that when the symbol or variable x is replaced by any object, say "a" from X, the statement p(a) is a proposition.

Example 1.3.1

Let X be the collection of all real numbers.

a. p(x): $x^2 + 1 \neq 0$
For $x = -1$, we have the proposition $(-1)^2 + 1 \neq 0$.
$x = 0$, we have the proposition $(0)^2 + 1 \neq 0$.
$x = 3$, we have the proposition $(3)^2 + 1 \neq 0$.

It should be noted that *for each* real number x, p(x) generates a *true* proposition.

b. p(x): $x^2 \leq x$
For $x = -\sqrt{2}$, we have the proposition $(-\sqrt{2})^2 \leq -\sqrt{2}$.
$x = \frac{1}{2}$, we have the proposition $(\frac{1}{2})^2 \leq \frac{1}{2}$.
$x = 3$, we have the proposition $(3)^2 \leq 3$.

It should be noted that *for some* real numbers x, p(x) generates a *true* proposition.

c. p(x): $x^2 + x + 6 = 0$
For $x = -2$, we have the proposition $(-2)^2 + (-2) + 6 = 0$.
$x = 1$, we have the proposition $(1)^2 + (1) + 6 = 0$.
$x = \sqrt{3}$, we have the proposition $(\sqrt{3})^2 + (\sqrt{3}) + 6 = 0$.

It should be noted that *for each* real number x, p(x) generates a *false* proposition.

If $p(x)$ is a formula defined on X, then the above examples show that $p(x)$ may be true for all objects of X, for some objects of X or no objects of X. Such phrases as "for all," "for each," "for every" are referred to as *universal quantifiers* and are denoted by the symbol "$\forall$." Phrases such as "for some," "for at least one," "there exists at least one" are referred to as *existential quantifiers* and are denoted by the symbol "$\exists$."

Let $p(x)$ be a formula defined on X. Then for each object of X, the formula $p(x)$ will generate a proposition.

The expression $(\forall x)(p(x))$ is to be read as: "For all objects x of X, $p(x)$," "For each object x of X, $p(x)$," or "For every object x of X, $p(x)$."

The expression $(\exists x)(p(x))$ is to be read as: "For some objects x of X, $p(x)$," "For at least one object x of X, $p(x)$," or "There exists at least one object x of X, $p(x)$."

Example 1.3.2

Let X be the collection of all real numbers.

a. If $p(x)$: $x^2 + 1 \neq 0$,

$(\forall x)(p(x))$ or $(\forall x)(x^2 + 1 \neq 0)$ reads "For each real number x, $x^2 + 1 \neq 0$."

$(\exists x)(p(x))$ or $(\exists x)(x^2 + 1 \neq 0)$ reads "For some real number x, $x^2 + 1 \neq 0$."

b. If $p(x)$: $x^2 \leq x$,

$(\forall x)(p(x))$ or $(\forall x)(x^2 \leq x)$ reads "For all real numbers , $x^2 \leq x$."

$(\exists x)(p(x))$ or $(\exists x)(x^2 \leq x)$ reads "For at least one real number x, $x^2 \leq x$."

If $p(x)$ is a formula defined on X, the expressions $(\forall x)(p(x))$ and $(\exists x)(p(x))$ may be thought of as propositions in themselves.

The proposition $(\forall x)(p(x))$ is true if and only if for each object of X, $p(x)$ is true. To establish that $(\forall x)(p(x))$ is a true proposition, one must show that for any arbitrary object x of X, $p(x)$ is true. To establish that $(\forall x)(p(x))$ is false, one must show that for at least one object x of X, $p(x)$ is false.

The proposition $(\exists x)(p(x))$ is true if and only if there is at least one object of X for which p(x) is true. To establish that $(\exists x)(p(x))$ is a true proposition, one must show that for some object x of X, p(x) is true. To establish that $(\exists x)(p(x))$ is false, one must show that there is no object x of X for which p(x) is true (or equivalently for all objects x of X, p(x) is false).

It becomes important to be able to establish the truth value of the propositions $(\forall x)(p(x))$ and $(\exists x)(p(x))$. The following example illustrates how this can be done.

Example 1.3.3

Let X be the collection of all real numbers.

a. If p(x): $x^2 + 1 \neq 0$, then p(x) is true for all real numbers. Hence, the proposition $(\forall x)(p(x))$ or $(\forall x)(x^2 + 1 \neq 0)$ is true.

   To establish that $(\forall x)(x^2 + 1 \neq 0)$ is true one must show that for any arbitrary real number x, p(x) is true. Let x be an arbitrary real number. Then $x^2 \geq 0$ and hence $x^2 + 1 \geq 1$. Therefore $x^2 + 1 \neq 0$ and p(x) is true.

b. If p(x): $x^2 \leq x$, then p(x) is true only for some real numbers. Hence the proposition $(\forall x)(p(x))$ or $(\forall x)(x^2 \leq x)$ is false, but the proposition $(\exists x)(p(x))$ or $(\exists x)(x^2 \leq x)$ is true.

   To establish that $(\forall x)(x^2 \leq x)$ is false, one must show that for some real number x, p(x) is false. In this case, when x = 3, p(3) is false since $9 \nleq 3$.

   To establish that $(\exists x)(x^2 \leq x)$ is true, one must show that for some real number x, p(x) is true. In this case, when $x = \frac{1}{2}$, $p(\frac{1}{2})$ is true since $\frac{1}{4} \leq \frac{1}{2}$.

c. If p(x): $x^2 + x + 6 = 0$, then p(x) is not true for all real numbers. Hence, the proposition $(\exists x)(p(x))$ or $(\exists x)(x^2 + x + 6 = 0)$ is false.

   To establish that $(\exists x)(x^2 + x + 6 = 0)$ is false, one must show that there is no real number x for which p(x) is true (or equivalently for all values of x, p(x) is false). Applying the quadratic formula to the equation $x^2 + x + 6 = 0$, one can show that there are no real solutions. Hence, there exists no real number x for which p(x) is true.

Since $(\forall x)(p(x))$ is a proposition, we can discuss its negation, $\sim [(\forall x)(p(x)]$. First, if $\sim [(\forall x)(p(x))]$ is true, then $(\forall x)(p(x))$ is false, i.e., it is not the case that for all objects x of X, p(x) is true. Hence for at least one object of X, p(x) is false and for that object $\sim p(x)$ is true. Therefore, $(\exists x)(\sim p(x))$ is true.

Secondly, if $\sim [(\forall x)(p(x))]$ is false, then $(\forall x)(p(x))$ is true, i.e., for all objects of X, p(x) is true. Hence for all objects of X, $\sim p(x)$ is false. Therefore $(\exists x)(\sim p(x))$ is false.

Thus we conclude $\sim [(\forall x)(p(x)]$ and $(\exists x)(\sim p(x))$ have the same truth tables and are logically equivalent. Similarly, $\sim [(\exists x)(p(x))]$ and $(\forall x)(\sim p(x))$ are logically equivalent.

Table 1.3.1

$$\sim [(\forall x)(p(x))] \equiv (\exists x)(\sim p(x))$$

$$\sim [(\exists x)(p(x))] \equiv (\forall x)(\sim p(x))$$

Example 1.3.4

Let X be the collection of all real numbers.

a. If p(x): $x^2 \leq x$, then the proposition $(\forall x)(x^2 \leq x)$ is false.

   There exists a real number x for which $x^2 \leq x$ is false; e.g., if $x = -2$, then $(-2)^2 \leq -2$ is false. Hence, the negation of $(\forall x)(x^2 \leq x)$ is true:

$$\begin{aligned} \sim [(\forall x)(x^2 \leq x)] &\equiv (\exists x)[\sim (x^2 \leq x)] \\ &\equiv (\exists x)(x^2 \nleq x) \end{aligned}$$

b. If p(x): $|x| = x$, then $(\exists x)(|x| = x)$ is true.

   There exists at least one real number x such that $|x| = x$ is true; e.g., if $x = 2$, then $|2| = 2$. Hence, the negation of $(\exists x)(|x| = x)$ is false:

$$\begin{aligned} \sim [(\exists x)(|x| = x)] &\equiv (\forall x)[\sim (|x| = x)] \\ &\equiv (\forall x)[(|x| \neq x)] \end{aligned}$$

c. If p(x): $x^2 + x + 6 = 0$, then the proposition $(\exists x)(x^2 + x + 6 = 0)$ is false.

   There exists no real number x such that $x^2 + x + 6 = 0$ since the quadratic equation has no real solutions. Hence the negation of $(\exists x)(x^2 + x + 6 = 0)$ is true:

$$\begin{aligned} \sim [(\exists x)(x^2 + x + 6 = 0)] &\equiv (\forall x)[\sim (x^2 + x + 6 = 0)] \\ &\equiv (\forall x)[(x^2 + x + 6 \neq 0)] \end{aligned}$$

Throughout our work in mathematics we shall often be required to show the proposition: $(\forall x)(p(x))$ is false. If $(\forall x)(p(x))$ is false, then $\sim[(\forall x)(p(x))]$ is true and hence $(\exists x)(\sim p(x))$ is true. This means that for at least one object x of X, $\sim p(x)$ is true. Thus p(x) is false for some object x of X. Consequently, to show the proposition $(\forall x)(p(x))$ is false, you must find one object of X for which p(x) is false. This is referred to as a *counterexample* to the proposition $(\forall x)(p(x))$. Hence, in Example 1.3.4, part (a), $x = -2$ is a counterexample to the statement $(\forall x)(x^2 \leq x)$.

Example 1.3.5

a. Let X be the collection of all men living in the U.S. and p(x): x wears green slacks. To show the proposition $(\forall x)(p(x))$, "All men in the U.S. wear green slacks" is false, you must find one man in the U.S. who does not wear green slacks. That man is a counterexample to "All men in the U.S. wear green slacks."

b. Let X be the collection of all real numbers. If p(x): $|x| = x$, the proposition $(\forall x)(|x| = x)$ is false since $|-3| \neq -3$. The value, $x = -3$, is a counterexample to $(\forall x)(|x| = x)$.

Exercises 1.3

For 1-4, let X be the collection of all real numbers. Determine the truth value of each of the propositions and then negate each proposition.

1. $(\forall x)(|x| > x)$

2. $(\exists x)(x^2 + x + 2 < 0)$

3. $(\forall x)(x^2 - 3x > 0)$

4. $(\exists x)(x^2 + x - 6 = 0)$

For 5 and 6, let X be the collection of all real numbers. By means of a counterexample show that each of the following propositions is false.

5. $(\forall x)(x^2 - 1 \geq 0)$

6. $(\forall x)$(If $|x| > 3$, then $x > 3$)

For 7-9, let **N** be the collection of all natural numbers. By means of a counterexample show each of the following propositions is false. (Note: A natural number n, $n > 1$, is prime if the only factors of n are 1 and n.)

7. $(\forall n)(n^2 + n + 17$ is prime)

8. $(\forall n)$(If n is prime, then n is odd)

9. $(\forall n)$(If $n \geq 100$, then n is not prime)

Section 1.4
ARGUMENTS AND PROOFS

At this point one might ask how logic fits into the scheme of things in mathematics. As mentioned earlier, logic plays an important role in the process of reasoning used in establishing the validity of statements referred to as "theorems." Mathematics is concerned with the study of what are called "mathematical systems." Essentially, a *mathematical system* consists of undefined or primitive terms, definitions, axioms and theorems.

Undefined or *primitive* terms in a given system are those basic concepts which are usually accepted intuitively in order to avoid circular definitions. (For example, in Euclidean plane geometry, "point" and "straight line" are usually considered as primitive terms.) *Defined terms* are those concepts whose meanings are explicitly stated in terms of the primitive ones. (For example, in Euclidean plane geometry, an "angle" is defined as the amount of opening between two straight lines which meet at a point.) *Axioms* are primary statements concerning one or more of the primitive terms. They specify the properties and relations that are assumed to be known about the primitive terms and are accepted without question as being true. (For example, in Euclidean plane geometry, we have the axiom: Through or connecting two given points, one and only one straight line can be drawn.) *Theorems* are statements which are the logical consequence of axioms, definitions or previously derived facts of the mathematical system under consideration. (For example, in Euclidean plane geometry we have the theorem: If two straight lines have two points in common, the lines coincide.)

Since theorems are statements which are not regarded as assumptions, the validity of such statements must be established. As we shall see in this section, the demonstration of the validity of a theorem requires both a knowledge of the rules of logic as well as the concept of a proof. In subsequent chapters we shall discuss from time to time how to approach mathematical proofs within the framework of various situations.

In this section we stress the logic of deductive proofs. Later, when more information is available, we shall convert to a more conventional narrative style of proof.

Definition 1.4.1

An ***argument*** is an assertion that a given sequence of propositions, $q_1, q_2, \ldots, q_n$, called the hypotheses (premises, assumptions), has as its consequence another proposition, q, called the conclusion.

Example 1.4.1

The following arguments are based on the law of detachment.

a. If it is raining, we shall not play golf. $\quad p \Rightarrow q$
It is raining. $\quad p$
∴ We shall not play golf. $\quad \therefore q$

b. It is raining or we shall play golf. $\quad p \vee q$
It is not raining. $\quad \sim p$
∴ We shall play golf. $\quad \therefore q$

Definition 1.4.2

An argument with hypotheses $q_1, q_2, \ldots, q_n$ and conclusion q is said to be a ***valid argument*** if q is always true whenever each of the hypotheses $q_1, q_2, \ldots, q_n$ is true.

Example 1.4.2

Consider the following arguments together with the accompanying tables.

a. $p \Rightarrow q$

$\underline{p}$

$\therefore q$

p	q	p ⇒ q
T	T	T
T	F	F
F	T	T
F	F	T

From the table it can be seen that the conclusion q is always true whenever the hypotheses $p \Rightarrow q$ *and* p are both true. Hence, the argument is valid.

b. $p \Rightarrow q$

$\underline{\sim p}$

$\therefore \sim q$

p	q	p ⇒ q	~ p	~ q
T	T	T	F	F
T	F	F	F	T
F	T	T	T	F
F	F	T	T	T

From the table it can be seen that in this example we have one instance where the conclusion ~q is false when the hypotheses are true. Thus the conclusion ~q is not always true whenever the hypotheses $p \Rightarrow q$ and ~p are both true. Hence, the argument is not valid.

It should be noted here that an argument with hypotheses $q_1, \ldots, q_n$ and conclusion q is valid if and only if the proposition $(q_1 \wedge \ldots \wedge q_n) \Rightarrow q$ is a tautology. To see this consider the following argument:

$$\begin{array}{c} q_1 \\ \cdot \\ \cdot \\ \cdot \\ \underline{q_n} \\ \therefore q \end{array}$$

First, suppose the argument is valid. We must show that $(q_1 \wedge \ldots \wedge q_n) \Rightarrow q$ is true regardless of the truth values of its components. If one of the q_i is false, then $q_1 \wedge \ldots \wedge q_n$ is false. Hence, by Definition 1.1.5, $(q_1 \wedge \ldots \wedge q_n) \Rightarrow q$ is true regardless of the truth value of q. If all the q_i are true; then, since the argument is valid, q is also true. Thus, by Definition 1.1.5, $(q_1 \wedge \ldots \wedge q_n) \Rightarrow q$ is true. Therefore, $(q_1 \wedge \ldots \wedge q_n) \Rightarrow q$ is a tautology.

Second, suppose the proposition $(q_1 \wedge \ldots \wedge q_n) \Rightarrow q$ is a tautology. We must show the argument is valid. Let $q_1, \ldots, q_n$ be true. Then $q_1 \wedge \ldots \wedge q_n$ is true. By Definition 1.1.5 (or equivalently by one of the laws of detachment), if $q_1 \wedge \ldots \wedge q_n$ is true and $(q_1 \wedge \ldots \wedge q_n) \Rightarrow q$ is true, then q must be true. Hence, the argument is valid.

Example 1.4.3

a.
$$\begin{array}{l} p \vee q \\ \sim p \\ \hline \therefore q \end{array}$$

The argument is valid since $[(p \vee q) \wedge (\sim p)] \Rightarrow q$ is a tautology.

b.
$$\begin{array}{l} p \Rightarrow q \\ \sim q \\ \hline \therefore \sim p \end{array}$$

The argument is valid since $[(p \Rightarrow q) \wedge (\sim q)] \Rightarrow \sim p$ is a tautology.

We need to develop methods whereby the validity of an argument can be readily established without the use of truth tables. Such a demonstration of the validity of an argument is called a *proof* and usually consists of a sequence of statements each of which is either a hypothesis of the argument, a tautology, a definition, a logically equivalent proposition (see Table 1.2.2) or a direct consequence of some preceding statement(s) derived at by one of the laws of logic (see Table 1.2.1). Since the procedure for constructing a proof is not necessarily unique, we shall investigate the various methods employed in the construction of a proof.

Consider an argument with hypotheses $q_1, q_2, \ldots, q_n$ and conclusion q. To show the argument is valid, we must show q is true whenever $q_1, q_2, \ldots, q_n$ are true or that $(q_1 \wedge q_2 \wedge \ldots \wedge q_n) \Rightarrow q$ is a tautology. A review of Table 1.2.1 and Table 1.2.2 might be helpful in following the steps of the various methods of a deductive proof given below.

> A ***Direct Proof*** of $(q_1 \wedge q_2 \wedge \ldots \wedge q_n) \Rightarrow q$ consists in assuming $q_1, q_2, \ldots, q_n$ are all true and then by using definitions, tautologies, logically equivalent statements and the various laws of logic one arrives at the truth of q. Since we establish that q is true whenever $q_1, q_2, \ldots, q_n$ are true, the argument is valid.

Example 1.4.4

Consider the following arguments and the proofs of their validity by the direct method. As the proof progresses, previous statements are referred to by their statement number.

a. p
$q \Rightarrow \sim p$
$\sim q \Rightarrow r$
$\therefore r$

proof:

1.	p	
2.	$q \Rightarrow \sim p$	
3.	$\sim q \Rightarrow r$	
4.	$\sim(\sim p) \Rightarrow \sim q$	Contrapositive Law: 2
5.	$p \Rightarrow \sim q$	Double Negation Law: 4
6.	$\sim q$	Detachment Law: 1, 5
7.	r	Detachment Law: 3, 6

■

b. $q \Rightarrow p$
$r \Rightarrow \sim p$
$s \vee r$
q
$\therefore s$

proof:

1.	$q \Rightarrow p$	
2.	$r \Rightarrow \sim p$	
3.	$s \vee r$	
4.	q	
5.	$\sim(\sim p) \Rightarrow \sim r$	Contrapositive Law: 2
6.	$p \Rightarrow \sim r$	Double Negation Law: 5
7.	$q \Rightarrow \sim r$	Transitive Law: 1, 6
8.	$\sim r$	Detachment Law: 4, 7
9.	s	Detachment Law: 3, 8

■

An ***Indirect Proof*** of $(q_1 \wedge q_2 \wedge \ldots \wedge q_n) \Rightarrow q$ consists in assuming $q_1, q_2, \ldots, q_n$ are all true and that the conclusion q is *false*; that is, one assumes $q_1, q_2, \ldots, q_n$ and $\sim q$ are all true. Then by using definitions, tautologies, logically equivalent statements and the various rules of logic one arrives at a proposition of the form $p \wedge \sim p$, which is a contradiction. This contradiction was arrived at by assuming the conclusion q was false. Hence, q must be true whenever $q_1, q_2, \ldots, q_n$ are all true. Therefore, the argument is valid.

Example 1.4.5

Consider the following arguments and the proofs of their validity by the indirect method.

a. p
$q \Rightarrow \sim p$
$\sim q \Rightarrow r$
$\therefore r$

proof:

1.	p	
2.	$q \Rightarrow \sim p$	
3.	$\sim q \Rightarrow r$	
4.	$\sim r$	
5.	$\sim r \Rightarrow \sim (\sim q)$	Contrapositive Law: 3
6.	$\sim r \Rightarrow q$	Double Negation Law: 5
7.	q	Detachment Law: 4, 6
8.	$\sim p$	Detachment Law: 2, 7
9.	$p \wedge \sim p$	Contradiction: 1, 8

$\therefore r$ ■

b. $r \Rightarrow s$
$q \vee r$
$\therefore q \vee s$

proof:

1.	$r \Rightarrow s$	
2.	$q \vee r$	
3.	$\sim (q \vee s)$	
4.	$\sim q \wedge \sim s$	DeMorgan's Law: 3
5.	$\sim q$	Simplification Law: 4
6.	$\sim s$	Simplification Law: 4
7.	$\sim r$	Detachment Law: 1, 6
8.	$\sim q \wedge \sim r$	Conjunction: 5, 7
9.	$\sim (q \vee r)$	DeMorgan's Law: 8
10.	$(q \vee r) \wedge \sim (q \vee r)$	Contradiction: 2, 9

$\therefore q \vee s$ ■

Since $(q_1 \wedge q_2 \wedge \ldots \wedge q_n) \Rightarrow q$ is logically equivalent to the proposition $\sim q \Rightarrow \sim (q_1 \wedge q_2 \wedge \ldots \wedge q_n)$, a variation of the above two methods of proof is referred to as the *contrapositive method.* Thus, if we assume $\sim q$ is true and show $\sim (q_1 \wedge q_2 \wedge \ldots \wedge q_n)$ is true, this is equivalent to proving $(q_1 \wedge q_2 \wedge \ldots \wedge q_n) \Rightarrow q$ and, therefore, the argument is valid.

Example 1.4.6

Consider the following argument and the proof of its validity by the contrapositive method:

$p \vee q$
$\sim p$
$\therefore q$

We wish to show that if $\sim q$ is true, then $\sim [(p \vee q) \wedge \sim p]$ is true. To see how the various steps of the proof below can be developed, let us find a logically equivalent proposition to $\sim [(p \vee q) \wedge \sim p]$:

$$\begin{aligned} \sim [(p \vee q) \wedge \sim p] &\equiv \sim (p \vee q) \vee \sim (\sim p) \\ &\equiv (\sim p \wedge \sim q) \vee p \\ &\equiv (\sim p \vee p) \wedge (\sim q \vee p) \end{aligned}$$

proof:

1.	$\sim q$	
2.	$\sim q \vee p$	Addition Law: 1
3.	$\sim p \vee p$	Tautology
4.	$(\sim p \vee p) \wedge (\sim q \vee p)$	Conjunction: 2, 3
5.	$(\sim p \wedge \sim q) \vee p$	Distributive Law: 4
6.	$\sim (p \vee q) \vee p$	DeMorgan's Law: 5
7.	$\sim [(p \vee q) \wedge \sim p]$	DeMorgan's Law: 6

■

Notice in the above example we started the thought process by working backwards until we could find a way to logically piece the steps of the proof together. Sometimes it is necessary to work in both directions before the framework of a proof becomes evident.

As your level of expertise in writing proofs increases, you will be tempted to delete "minor" steps in your proofs. This is natural; and, in fact, desirable since it will be impossible to include explanations for every step in a proof. The difficulty of deciding which steps can be omitted and which steps should not be omitted will become easier after you have read and done many proofs.

Exercises 1.4

For 1 and 2, an argument and the proof of its validity have been furnished. Indicate the method of proof being used and give a reason for each step of the proof.

1. $p \Rightarrow (q \Rightarrow r)$
   $p \wedge q$
   ______
   $\therefore r$

proof:

1.	$p \Rightarrow (q \Rightarrow r)$	1.	________
2.	$p \wedge q$	2.	________
3.	p	3.	________
4.	$q \Rightarrow r$	4.	________
5.	q	5.	________
6.	r	6.	________

2. $\sim p \Rightarrow \sim q$
   $\sim r$
   $p \Rightarrow s$
   $q \vee r$
   ______
   $\therefore s$

proof:

1.	$\sim p \Rightarrow \sim q$	1.	________
2.	$\sim r$	2.	________
3.	$p \Rightarrow s$	3.	________
4.	$q \vee r$	4.	________
5.	$\sim s$	5.	________
6.	$\sim s \Rightarrow \sim p$	6.	________
7.	$\sim p$	7.	________
8.	$\sim q$	8.	________
9.	$\sim q \wedge \sim r$	9.	________
10.	$\sim (q \vee r)$	10.	________
11.	$(q \vee r) \wedge \sim (q \vee r)$	11.	________

$\therefore s$

For 3-6, provide proofs for the following valid arguments. Indicate the method of proof being used and give a reason for each step of the proof.

3. $p \vee q$
   $\sim q \vee r$
   ______
   $\therefore p \vee r$

4. $p \vee q \Rightarrow p \wedge r$
   $\sim p$
   ______
   $\therefore \sim q$

5. q
   $p \Rightarrow \sim q$
   $r \Rightarrow p$
   ______
   $\therefore \sim r$

6. $p \vee (q \wedge r)$
   $q \Rightarrow s$
   $r \Rightarrow t$
   $(s \wedge t) \Rightarrow w$
   $\sim p$
   ______
   $\therefore w$

Section 1.5
MATHEMATICAL INDUCTION

The proofs discussed in the previous section are referred to as *deductive*. Such proofs were constructed by using the rules of logic to draw a particular conclusion based on statements (hypotheses) accepted as being true.

Another method is called *inductive*. In an inductive process, one forms a general conclusion on the basis of looking at particular cases. In this section we formalize the inductive process to a method of proof.

In many instances an argument involves a statement about the natural numbers. To establish the validity of such an argument we use the Principle of Mathematical Induction. There are two versions of this principle which we shall consider.

Mathematical Induction I

Let p(n) be a formula defined on the natural numbers 1, 2, 3, 4, . . . such that the following two conditions are satisfied:

a. p(1) is true
b. Whenever p(k) is true for any natural number k, then p(k + 1) is also true.

Then p(n) generates a true proposition for every natural number n; i.e., $(\forall n)(p(n))$ is true.

To show the reasonableness of this principle, assume that conditions (a) and (b) are satisfied. Let us now select an arbitrary natural number, say t. We wish to show that p(t) is true. From condition (a), we know p(1) is true. Hence, by the repeated application of condition (b), we have

$$\begin{aligned} p(1+1) &= p(2) \text{ is true.} \\ p(2+1) &= p(3) \text{ is true.} \\ &\vdots \\ p([t-2]+1) &= p(t-1) \text{ is true.} \\ p([t-1]+1) &= p(t) \text{ is true.} \end{aligned}$$

Since t is an arbitrary natural number, we conclude that $(\forall n)(p(n))$ is true.

Example 1.5.1

Prove each of the following statements is true using the Principle of Mathematical Induction I.

a. $1 + 2 + \ldots + n = \frac{n(n+1)}{2}$

proof: Let p(n): $1 + 2 + \ldots + n = \frac{n(n+1)}{2}$

p(1): Show p(1) is true; i.e., $1 = \frac{1(1+1)}{2}$.

$$\frac{1(1+1)}{2} = \frac{1(2)}{2} = 1$$

$\therefore$ p(1) is true.

p(k): Assume p(k) is true for some k, $k \geq 1$.

Then $1 + 2 + \ldots + k = \frac{k(k+1)}{2}$.

p(k + 1): Show p(k + 1) true; i.e.,

$$1 + 2 + \ldots + k + (k+1) = \frac{(k+1)[(k+1)+1]}{2}.$$

$$\begin{aligned} 1 + 2 + \ldots + k + (k+1) &= \frac{k(k+1)}{2} + (k+1) \\ &= \frac{k(k+1) + 2(k+1)}{2} \\ &= \frac{(k+1)(k+2)}{2} \\ &= \frac{(k+1)[(k+1)+1]}{2} \end{aligned}$$

$\therefore$ p(k + 1) is true.

Thus $(\forall n)(p(n))$ is true. ■

Example 1.5.1 continued next page.

Example 1.5.1 Continued

b. $1 + 3 + 5 + \ldots + (2n - 1) = n^2$

proof: Let p(n): $1 + 3 + 5 + \ldots + (2n - 1) = n^2$

p(1): Show p(1) is true; i.e., $1 = (1)^2$.

$(1)^2 = 1$
$\therefore$ p(1) is true.

p(k): Assume p(k) is true for some k, $k \geq 1$.
Then $1 + 3 + 5 + \ldots + (2k - 1) = k^2$.

p(k + 1): Show p(k + 1) is true, i.e.,
$1 + 3 + 5 + \ldots + (2k - 1) + [2(k + 1) - 1] = (k + 1)^2$.

$$\begin{aligned} 1 + 3 + 5 + \ldots + (2k - 1) + (2k + 1) &= k^2 + (2k + 1) \\ &= k^2 + 2k + 1 \\ &= (k + 1)^2 \end{aligned}$$

$\therefore$ p(k + 1) is true.

Thus $(\forall n)(p(n))$ is true. ■

c. **(Laws of Exponents)** Recall that, if a is a real number and n a natural number, we define $a^1 = a$, $a^2 = a \cdot a, \ldots, a^n = a \cdot a \cdots a$ (n times) and $a^{n+1} = a^n a$. Using Mathematical Induction I, we shall show the following rules are true:

i. If a and b are real numbers, then $(ab)^n = a^n b^n$

proof: Let p(n): $(ab)^n = a^n b^n$

p(1): Show p(1) is true; i.e., $(ab)^1 = a^1 b^1$.

$$\begin{aligned} (ab)^1 &= ab \\ &= a^1 b^1 \end{aligned}$$
$\therefore$ p(1) is true.

p(k): Assume p(k) is true for some k, $k \geq 1$.
Then $(ab)^k = a^k b^k$.

p(k + 1): Show p(k + 1) is true; i. e., $(ab)^{k+1} = a^{k+1}b^{k+1}$.

$(ab)^{k+1}$	$= (ab)^k(ab)$	Definition
	$= (a^kb^k)(ab)$	p(k) is true
	$= (a^ka)(b^kb)$	Associative/Commutative
	$= (a^{k+1})(b^{k+1})$	Definition

$\therefore$ p(k + 1) is true.

Thus $(\forall n)(p(n))$ is true. ■

ii. If a is a real number and m and n are natural numbers, then $a^ma^n = a^{m+n}$.

proof: We must show that $a^ma^n = a^{m+n}$ is true for all natural numbers m and n. Let m be an arbitrary fixed natural number and let p(n): $a^ma^n = a^{m+n}$.

p(1): Show p(1) is true; i. e., $a^ma^1 = a^{m+1}$.

a^ma^1	$= a^ma$	Definition
	$= a^{m+1}$	Definition

$\therefore$ p(1) is true.

p(k): Assume p(k) is true for some k, $k \geq 1$. Then $a^ma^k = a^{m+k}$.

p(k + 1): Show p(k + 1) is true; i.e., $a^ma^{k+1} = a^{m+(k+1)}$.

a^ma^{k+1}	$= a^m(a^ka)$	Definition
	$= (a^ma^k)a$	Associative
	$= a^{m+k}a$	p(k) is true
	$= a^{(m+k)+1}$	Definition
	$= a^{m+(k+1)}$	Associative

$\therefore$ p(k + 1) is true.

Thus we have $(\forall n)(p(n))$ is true. Since m is an arbitrary natural number, $a^ma^n = a^{m+n}$ is true for all natural numbers m and n, and we write $(\forall m,n)(a^ma^n = a^{m+n})$. ■

Example 1.5.1 continued next page.

Example 1.5.1 Continued

d. $n < 2^n$

proof: Let p(n): $n < 2^n$

p(1): Show p(1) is true; i.e., $1 < 2^1$.

Since $1 < 2$ and $2^1 = 2$, $1 < 2^1$
$\therefore$ p(1) is true.

p(k): Assume p(k) is true for some k, $k \geq 1$.
Then $k < 2^k$.

p(k + 1): Show p(k + 1)is true; i.e., $(k + 1) < 2^{k+1}$.

Now
$$k < 2^k$$
$$2(k) < 2(2^k)$$
$$2k < 2^{k+1}$$

Since
$$1 \leq k$$
$$k + 1 \leq k + k$$
$$k + 1 \leq 2k$$

Therefore, we have $k + 1 \leq 2k$ and $2k < 2^{k+1}$. By the transitive rule, $k + 1 < 2^{k+1}$.
$\therefore$ p(k + 1) is true.

Thus $(\forall n)(p(n))$ is true. ■

The basic Principle of Mathematical Induction I can be extended to a more general principle. Let a be a *fixed* natural number and p(n) be a formula defined on the natural numbers a, a + 1, a + 2, . . . such that the following two conditions are satisfied:

a. p(a) is true.

b. Whenever p(k) is true for any natural number k, $k \geq a$, then p(k + 1) is also true.

Then p(n) generates a true proposition for every natural number n where $n \geq a$; i.e., $(\forall n, n \geq a)(p(n))$ is true.

Example 1.5.2

a. (**General Transitive Law**) Recall that if a, b and c are real numbers such that $a < b$ and $b < c$, then by the transitive law for inequalities $a < c$. Using the more general version of Mathematical Induction I, let us show that for $n \geq 3$, if $a_1, \ldots, a_n$ are real numbers such that

$$a_1 < a_2,\ a_2 < a_3, \ldots, a_{n-1} < a_n, \text{ then } a_1 < a_n.$$

proof: For $n \geq 3$, let p(n) be the statement:

If $a_1 < a_2$, $a_2 < a_3, \ldots, a_{n-1} < a_n$, then $a_1 < a_n$.

p(3): Show p(3) is true; i.e., if $a_1 < a_2$ and $a_2 < a_3$, then $a_1 < a_3$.
This is true by the transitive law.
∴ p(3) is true.

p(k): Assume p(k) is true for some k, $k \geq 3$.
Thus if $a_1 < a_2$, $a_2 < a_3, \ldots, a_{k-1} < a_k$, then $a_1 < a_k$.

p(k + 1): Show p(k + 1) is true; i.e., if
$a_1 < a_2, \ldots, a_{k-1} < a_k$, $a_k < a_{k+1}$, then $a_1 < a_{k+1}$.

Let $a_1 < a_2, \ldots, a_{k-1} < a_k$, $a_k < a_{k+1}$.
Since $a_1 < a_2, \ldots, a_{k-1} < a_k$ and p(k) is true, $a_1 < a_k$.
We now have $a_1 < a_k$ and $a_k < a_{k+1}$.
By the transitive law, $a_1 < a_{k+1}$.
∴ p(k + 1) is true.

Thus $(\forall n \geq 3)(p(n))$ is true. ■

Example 1.5.2 continued next page.

Example 1.5.2 Continued

b. **(General Distributive Law)** Recall that if a, b and c are real numbers then by the distributive law $a(b + c) = ab + ac$. Using the more general version of Mathematical Induction I, let us show that for $n \geq 2$, if $a, a_1, a_2, \ldots, a_n$ are real numbers, then $a(a_1 + a_2 + \ldots + a_n) = a\,a_1 + a\,a_2 + \ldots + a\,a_n$.

proof: For $n \geq 2$, let p(n) be the statement:

$$a(a_1 + a_2 + \ldots + a_n) = a\,a_1 + a\,a_2 + \ldots + a\,a_n.$$

p(2): Show p(2) is true; i.e., $a(a_1 + a_2) = a\,a_1 + a\,a_2$.
This is true by the distributive law.
$\therefore$ p(2) is true.

p(k): Assume p(k) is true for some k, $k \geq 2$.
Then $a(a_1 + a_2 + \ldots + a_k) = a\,a_1 + a\,a_2 + \ldots + a\,a_k$.

p(k + 1): Show p(k + 1) is true;
i.e., $a(a_1 + \ldots + a_k + a_{k+1}) = a\,a_1 + \ldots + a\,a_k + a\,a_{k+1}$.

$$\begin{aligned} a(a_1 + a_2 + \ldots + a_k + a_{k+1}) &= a[(a_1 + a_2 + \ldots + a_k) + a_{k+1}] \\ &= a[b + a_{k+1}] \text{ where } b = a_1 + a_2 + \ldots + a_k \\ &= a\,b + a\,a_{k+1} \text{ by the distributive law} \\ &= a(a_1 + a_2 + \ldots + a_k) + a\,a_{k+1} \\ &= (a\,a_1 + a\,a_2 + \ldots + a\,a_k) + a\,a_{k+1} \\ &= a\,a_1 + a\,a_2 + \ldots + a\,a_k + a\,a_{k+1} \end{aligned}$$

$\therefore$ p(k + 1) is true.

Thus, $(\forall n \geq 2)(p(n))$ is true. ■

Another form of mathematical induction is given by the following principle.

Mathematical Induction II

Let a be a *fixed* natural number and p(n) a formula defined on the natural numbers a, a + 1, a + 2, . . . such that the following conditions are satisfied:

a. p(a) is true.

b. Whenever p(k) is true for any natural number k, where $a \leq k < m$, then p(m) is also true.

Then p(n) generates a true proposition for every natural number n where $n \geq a$; i.e., $(\forall n, n \geq a)(p(n))$ is true.

To show the reasonableness of the principle, assume that conditions (a) and (b) hold and that p(n) is *not* true for all natural numbers n, where $n \geq a$. Since p(n) is not true for all n, $n \geq a$, $(\forall n, n \geq a)(p(n))$ is false. Therefore, $\sim[(\forall n, n \geq a)(p(n))] \equiv (\exists n, n \geq a)(\sim p(n))$ is true. Hence, there exists a natural number t such that $t \geq a$ and $\sim p(t)$ is true or p(t) is false. Thus there is a smallest natural number m, $m \geq a$, for which p(m) is false. Consequently, p(k) is true for all k, $a \leq k < m$. By condition (b), p(m) must be true. We therefore have $p(m) \wedge \sim p(m)$ which is a contradiction. Thus p(n) generates a true proposition for every natural number n, $n \geq a$.

Example 1.5.3

a. (Example 1.5.1, b, revisited).

proof: For $n \geq 1$, let p(n): $1 + 3 + 5 + \ldots + (2n - 1) = n^2$.

p(1): Show p(1) is true; i.e., $1 = 1^2$.

$(1)^2 = 1$
$\therefore$ p(1) is true.

p(k): Assume p(k) is true for all k, $1 \leq k < m$.
Then for any k, $1 \leq k < m$, $1 + \ldots + (2k - 1) = k^2$.

p(m): Show p(m) is true; i.e., $1 + \ldots + (2m - 1) = m^2$.

Let $k = m - 1$. Since $1 \leq m - 1 < m$, p(m – 1) is true;
i.e., $1 + 3 + 5 + \ldots + [2(m - 1) - 1] = (m - 1)^2$.

Hence,

$$\begin{aligned} 1 + \ldots + [2(m-1) - 1] + (2m - 1) &= (m-1)^2 + (2m - 1) \\ &= m^2 - 2m + 1 + 2m - 1 \\ &= m^2 \end{aligned}$$

$\therefore$ p(m) is true.

Thus $(\forall n)(p(n))$ is true. ■

Example 1.5.3 continued next page.

Example 1.5.3 Continued

b. A natural number p, $p > 1$, is said to be *prime* if the only factors of p are 1 and p. Thus 2, 3, 5, 7, 11, etc., are prime natural numbers. We wish to show that every natural number n, $n \geq 2$, has a prime factor.

proof: For $n \geq 2$, let p(n): n has a prime factor.

p(2): Show p(2) is true; i.e., 2 has a prime factor.

Now 2 is prime and 2 is a factor of 2.
$\therefore$ p(2) is true.

p(k): Assume p(k) is true for all k, $1 \leq k < m$. Then for any k, $1 \leq k < m$, k has a prime factor.

p(m): Show p(m) is true; i.e., m has a prime factor.

If m is prime, then p(m) is true since m is a factor of m. If m is not prime, then m has a factor t where $t \neq 1, m$. Hence $m = ts$ where $1 < t < m$. Since $1 < t < m$, p(t) is true, i.e., t has a prime factor p. Let $t = pq$. Then $m = ts = pqs$ and m has a prime factor.
$\therefore$ p(m) is true.

$\therefore$ $(\forall n, n \geq 2)(p(n))$ is true. ■

Exercises 1.5

For 1-6, prove each by Mathematical Induction.

1. $1^2 + 2^2 + \ldots + n^2 = \frac{n(n+1)(2n+1)}{6}$

2. $2^0 + 2^1 + \ldots + 2^{n-1} = 2^n - 1$

3. $1 \cdot 2 + 2 \cdot 3 + 3 \cdot 4 + \ldots + n(n+1) = \frac{n(n+1)(n+2)}{3}$

4. $1^2 + 4^2 + 7^2 + \ldots + (3n-2)^2 = \frac{n(6n^2 - 3n - 1)}{2}$

5. Let a be a real number such that $a > 0$.
   Prove: $(a+1)^n > na + 1$ where $n \geq 2$.

6. Let a be a real number and m and n natural numbers.
   Prove the following law for exponents: $(a^m)^n = a^{mn}$

For 7-9, recall that, if a is a real number and n a natural number, we define $1a = a$, $2a = a + a$, . . ., $na = a + a + \ldots + a$ (n times) and $(n+1)a = na + a$. Using Mathematical Induction, prove each of the following.

7. Let a and b be real numbers. Then $n(a+b) = na + nb$.

8. Let a be a real number and m and n natural numbers.
   Then $(m+n)a = ma + na$.

9. Let a be a real number and m and n natural numbers.
   Then $(mn)a = m(na)$.

10. Let a, b and c be real numbers and define $a + b + c = (a + b) + c$.
    For $n \geq 3$, let p(n) be the statement:
    $$a_1 + a_2 + \ldots + a_{n-1} + a_n = (a_1 + a_2 + \ldots + a_{n-1}) + a_n.$$
    Using Mathematical Induction, show $(\forall n, n \geq 3)(p(n))$ is true.

Chapter 2

ALGEBRA OF SETS

Almost all branches of mathematics require a working knowledge of the algebra of sets. In this chapter we deal with the basic concepts of the algebra of sets which will provide a common thread to those branches that utilize it. Additionally, we shall attempt to provide you with techniques that will assist you in developing a more common narrative style of proof.

Section 2.1
SETS

You probably have a general idea as to the meaning of a set or collection of objects. You might think of a set of dishes, a deck of cards or an Admissions Committee. Unfortunately, it is technically impossible to define a "set" in a precise way except by the utilization of other undefined or primitive terms.

We shall adopt the convention that a *set* may be thought of intuitively as a well-defined collection of objects. Well-defined means that given any object and a set, one can determine with certainty whether or not the object belongs to that set. The objects organized into a set are called the *elements* or *members* of the set.

Example 2.1.1

The following are examples of sets:

a. The set of all counties in the state of New York.

b. The set of all state capitals in the United States.

c. The set of all integers divisible by 3.

A set may be "finite" or "infinite" (see Chapter 7) depending on the number of elements it contains. If a set contains no elements, it is called the *empty* set or *null* set and denoted by $\emptyset$ (phi).

In most cases, sets will be designated by the use of upper-case letters such as A, B, etc., and elements by the use of lower-case letters such as a, b, etc. To indicate that the object x is or is not an element of the set A, the conventional notation is:

$x \in A$ if x is an element of set A; *and*
$x \notin A$ if x is not an element of set A.

Note that the statements $x \notin A$ and $\sim(x \in A)$ are logically equivalent by observing that $x \notin A \Leftrightarrow \sim(x \in A)$ is a tautology.

$x \in A$	$x \notin A$	$\sim(x \in A)$	$x \notin A \Leftrightarrow \sim(x \in A)$
T	F	F	T
F	T	T	T

Example 2.1.2

a. If A is the set of all state capitals in the United States, then A is a "finite" set and Albany $\in$ A while Chicago $\notin$ A.

b. If B is the set of all integers divisible by 3, then B is an "infinite" set and -12 $\in$ B while 25 $\notin$ B.

c. If C is the set of all real numbers for which $x^2 + x + 6 = 0$, then C has no elements; i.e., C = $\emptyset$.

Definition 2.1.1

Two sets, A and B, are said to be ***equal*** and we write **A = B** if A and B have the same elements; that is, every element of A is an element of B and every element of B is an element of A. Symbolically,

$$(A = B) \Leftrightarrow (\forall x)(x \in A \Leftrightarrow x \in B)$$

It is often necessary to exhibit the elements of a set. This is done by enclosing the elements of the set with braces, { }. For example, if A is the set whose elements are x, y and z, we write A = {x,y,z}.

If x is an element of a set, x and {x} are considered differently. {x} denotes the set containing the element x, whereas x is the element of the set {x}; i.e., x $\in$ {x}.

There are two direct consequences of the definition of set equality. First, unless specified, the order in which elements appear in a given set is immaterial. Thus, {a,b,c} = {a,c,b}. Second, the elements in a set are usually listed without duplication. Thus, {a,b,a,c} = {a,b,c}.

Throughout the text it will be convenient to use standard symbols for some common sets. Hence,

N = the set of natural numbers
= {1,2,3, . . .};

Z = the set of integers
= { . . . ,-3,-2,-1,0,1,2,3, . . . };

Q = the set of rational numbers;

R = the set of real numbers.

Exercises 2.1

For 1-7, list the elements of each of the following sets described in words.

1. The set of integers between 6 and 13 inclusive.

2. The set of integers between 6 and 8.

3. The set of integers between 1 and 2.

4. The set of negative integers.

5. The set of integers which have 3 as a factor.

6. The set of all integers which are neither positive nor negative.

7. The set of odd positive integers.

Section 2.2
SUBSETS

We have already discussed the equality of sets. It may be that every element in one set, say A, is an element of another set, say B, but the converse may not be true; that is, some elements in set B may not be elements of set A. In this case we say that the first set, A, is a *subset* of the second, B.

Definition 2.2.1

A set A is a ***subset*** of a set B and we write $\mathbf{A \subseteq B}$ or $\mathbf{B \supseteq A}$ if and only if each element of A is an element of B. If A is a subset of B, then B is called a ***superset*** of A. Symbolically,

$$(A \subseteq B) \Leftrightarrow (\forall x)(x \in A \Rightarrow x \in B)$$

How does one go about showing $A \subseteq B$? The definition of subset tells us that we must show every element of set A is an element of set B. To prove a statement involving the universal quantifier, one selects an arbitrary element $x \in A$. Using the fact that $x \in A$, one must show that $x \in B$. Having demonstrated this, one may conclude $(\forall x)(x \in A \Rightarrow x \in B)$; i.e., $A \subseteq B$, since x is an arbitrary element of A.

How does one go about showing $A = B$? By the definition of set equality, $A = B$ if $(\forall x)(x \in A \Leftrightarrow x \in B)$. That is, $A = B$ if every element of A is an element of B; i.e., $(\forall x)(x \in A \Rightarrow x \in B)$ *and* if every element of B is an element of A; i.e., $(\forall x)(x \in B \Rightarrow x \in A)$. Hence, to show $A = B$ one must show $A \subseteq B$ *and* $B \subseteq A$.

If A is not a subset of B, we write $A \not\subseteq B$. Therefore, if $A \not\subseteq B$, then $(\forall x)(x \in A \Rightarrow x \in B)$ is false. Thus $\sim[(\forall x)(x \in A \Rightarrow x \in B)]$ is true. But $\sim[(\forall x)(x \in A \Rightarrow x \in B)] \equiv (\exists x)[\sim(x \in A \Rightarrow x \in B)]$. From Table 1.2.2, $\sim(x \in A \Rightarrow x \in B) \equiv \sim[\sim(x \in A) \vee (x \in B)] \equiv [(x \in A) \wedge \sim(x \in B)] \equiv (x \in A \wedge x \notin B)$. Consequently, $\sim[(\forall x)(x \in A \Rightarrow x \in B)] \equiv (\exists x)(x \in A \wedge x \notin B)$. Hence, to show $A \not\subseteq B$ one must demonstrate that there exists at least one element x such that $x \in A$ and $x \notin B$.

Example 2.2.1

a. Let $A = \{1,2,5\}$, $B = \{1,2,\ldots,10\}$ and $C = \{0,1,2\}$. $A \subseteq B$, since each element of A is also an element of B, while $C \not\subseteq B$ since $0 \in C$ but $0 \notin B$.

b. Given the sets: **N**, **Z**, **Q**, **R** mentioned earlier; $\mathbf{N} \subseteq \mathbf{Z}$ since each natural number is an integer while $\mathbf{R} \not\subseteq \mathbf{Q}$ since $\sqrt{2}$ is a real number but $\sqrt{2}$ is not a rational number (see Chapter 8).

c. If $A = \{a,b,c,d\}$ and $B = \{b,d,a,c\}$, then $A = B$ since $A \subseteq B$ and $B \subseteq A$.

Theorem 2.2.1

Let A be any set. Then,

a. $A \subseteq A$ b. $\emptyset \subseteq A$

Proof

Since the proof of part (a) is obvious, we shall prove part (b).

b. Let A be a set. We shall show $x \in \emptyset \Rightarrow x \in A$ is a tautology. Since the statement $x \in \emptyset$ is false, we have the following truth table for the proposition $x \in \emptyset \Rightarrow x \in A$.

$x \in \emptyset$	$x \in A$	$x \in \emptyset \Rightarrow x \in A$
F	T	T
F	F	T

Hence, $(\forall x)(x \in \emptyset \Rightarrow x \in A)$ and we have $\emptyset \subseteq A$. ■

Theorem 2.2.2

Let A, B and C be sets.

a. If $A \subseteq B$ and $B \subseteq C$, then $A \subseteq C$.

b. If $A \subseteq B$ and $x \notin B$, then $x \notin A$.

Proof

a. Let $A \subseteq B$ and $B \subseteq C$ (hypotheses). We wish to show $A \subseteq C$; i.e., $(\forall x)(x \in A \Rightarrow x \in C)$. Hence, select an arbitrary element $x \in A$.

$x \in A$	
$x \in A \Rightarrow x \in B$	$A \subseteq B$
$x \in B$	Detachment Law
$x \in B \Rightarrow x \in C$	$B \subseteq C$
$x \in C$	Detachment Law

Therefore, $A \subseteq C$. ■

b. Let $A \subseteq B$ and $x \notin B$ (hypotheses). We wish to show that $x \notin A$. Using the indirect method of proof, assume $x \notin A$ is false; i.e., that $\sim(x \notin A) \equiv x \in A$ is true.

$x \in A$	
$x \in A \Rightarrow x \in B$	$A \subseteq B$
$x \in B$	Detachment Law

Hence, we have $(x \notin B) \wedge (x \in B)$ which is a contradiction. The contradiction arose by assuming $x \notin A$ was false. Therefore, $x \notin A$ is true. ■

When $A \subseteq B$ and $A \neq B$, A is called a *proper subset* of B and we write **$A \subset B$** or **$B \supset A$**. Now, if $A \subset B$, then $A \subseteq B$ but $B \nsubseteq A$. Hence, A is a proper subset of B if every element of A is an element of B and if there exists an element of B which is not an element of A. Thus, to show $A \subset B$ one must show:

$(\forall x)\ (x \in A \Rightarrow x \in B)$; *and*
$(\exists x)(x \in B \wedge x \notin A)$.

Example 2.2.2

a. $\mathbf{N} \subset \mathbf{Z}$ since $\mathbf{N} \subseteq \mathbf{Z}$ and $-2 \in \mathbf{Z}$, but $-2 \notin \mathbf{N}$.

b. $\mathbf{Z} \subset \mathbf{Q}$ since $\mathbf{Z} \subseteq \mathbf{Q}$ and $\frac{1}{2} \in \mathbf{Q}$, but $\frac{1}{2} \notin \mathbf{Z}$.

c. $\mathbf{Q} \subset \mathbf{R}$ since $\mathbf{Q} \subseteq \mathbf{R}$ and $\sqrt{2} \in \mathbf{R}$, but $\sqrt{2} \notin \mathbf{Q}$.

In addition to listing elements to specify a given set, a set may be described by *set builder* notation. When using this notation, it is necessary to have some superset, say X, from which the set is to be built. We define a formula p(x) on X such that for each $x \in X$, p(x) becomes a proposition. In this way, we can construct a set which contains those elements $x \in X$ for which p(x) is true. The notation used is:

$$\{x \in X \mid p(x)\}$$

which one reads as "the set of $x \in X$ such that p(x) is true." For example, if $X = \mathbf{Z}$ and p(x): x is an even integer, then the set

$E = \{x \in \mathbf{Z} \mid p(x)\} = \{x \in \mathbf{Z} \mid x \text{ is an even integer}\}$.

Example 2.2.3

a. Describe the following by set-builder notation:

i. $A = \{1,3,5,7,\ldots\}$. Then A is the set of odd natural numbers. That is, $A = \{x \in \mathbf{N} \mid x = 2n - 1, n \in \mathbf{N}\}$.

ii. $B = \{-2,4,-8,16,-32,\ldots\}$. Then B is the set of all integers that can be expressed as a positive power of -2. That is, $B = \{(-2)^1,(-2)^2,(-2)^3,(-2)^4,(-2)^5,\ldots\} = \{x \in \mathbf{Z} \mid x = (-2)^n, n \in \mathbf{N}\}$.

Example 2.2.3 continued next page.

Example 2.2.3, Continued

b. For each of the following, list the elements of the set:

   i. Let $C = \{x \in \mathbf{N} \mid -1 < x < 3\}$.
   If $x \in C$, then $x \in \mathbf{N}$ and $-1 < x < 3$.
   Therefore, $C = \{1,2\}$.

   ii. Let $D = \{x \in \mathbf{R} \mid x^2 < x\}$.
   If $x \in D$, then $x \in \mathbf{R}$ and $x^2 < x$.
   Now, $x^2 < x \Rightarrow x^2 - x < 0 \Rightarrow x(x - 1) < 0 \Rightarrow 0 < x < 1$.
   Therefore, $D = \{x \in R \mid 0 < x < 1\}$.

   iii. If $E = \{x \in \mathbf{R} \mid x^2 + 1 = 0\}$.
   If $x \in E$, then $x \in \mathbf{R}$ and $x^2 + 1 = 0$.
   Now, for each real number x, $x^2 \geq 0 \Rightarrow x^2 + 1 \geq 1$.
   Therefore, for each real number $x^2 + 1 \neq 0$ and hence $E = \emptyset$.

For every set A, the set of all subsets of A, denoted by **P(A)**, is called the *power set of A*; i.e.,

$$P(A) = \{B \mid B \subseteq A\}.$$

P(A) represents a set containing sets as its elements. After the next example, we should be able to make a conjecture as to the number of elements in the power set of A if A is a finite set.

Example 2.2.4

a. Find P(A) if $A = \emptyset$. $P(\emptyset) = \{\emptyset\}$. Note that $P(\emptyset) \neq \emptyset$ since $\emptyset \in P(\emptyset)$.

b. Find P(A) if $A = \{a\}$. $P(A) = \{\emptyset, A\}$.

c. Find P(A) if $A = \{a,b\}$. $P(A) = \{\emptyset, \{a\}, \{b\}, A\}$.

d. Find P(A) if $A = \{a,b,c\}$. $P(A) = \{\emptyset,\{a\},\{b\},\{c\},\{a,b\},\{a,c\},\{b,c\}, A\}$.

Parts (a) - (d) of Example 2.2.4 would seem to indicate that if a set A contains n elements, then its power set, P(A), contains 2^n elements. This conjecture is indeed true. (See Exercises 3.2, parts (6) - (9).)

Exercises 2.2

1. List the elements of the sets given by set-builder notation:

   a. $A = \{x \in \mathbf{Z} \mid x^2 > 25\}$

   b. $B = \{x \in \mathbf{N} \mid x = n^3 + 1, n \in \mathbf{N}\}$

   c. $C = \{x \in \mathbf{Z} \mid -6 \leq x \leq 2\}$

   d. $D = \{x \in \mathbf{Z} \mid \sqrt{x} \in \mathbf{Z}, x \leq 144\}$

2. Describe the following sets by set-builder notation:

   a. $C = \{0, \frac{1}{2}, \frac{2}{3}, \frac{3}{4}, \ldots\}$

   b. $D = \{a,e,i,o,u\}$

   c. $E = \{1,2, \ldots ,100\}$

   d. $F = \{0,3,6,9, \ldots\}$

   e. $G = \{1,4,9,16,25,36,49,64,81,100\}$

3. By means of a counterexample show each of the following statements is false.

   a. If $A \subseteq B$ and $B \not\subseteq C$, then $A \not\subseteq C$.

   b. If $A \not\subseteq B$ and $B \not\subseteq C$, then $A \not\subseteq C$.

   c. If $x \in A$ and $A \not\subseteq B$, then $x \notin B$.

4. Find $P(A)$ if $A = \{a,b,c,d\}$.

Section 2.3
SET OPERATORS

The basic operations utilized in the arithmetic of real numbers are addition, subtraction, multiplication and division. Given two numbers, a and b, there is assigned an unique number $a + b$, $a - b$, ab or $a \div b$ ($b \neq 0$) according to specified rules. In like manner, for a given set X, one can define operations on the power set P(X) — the *union*, the *difference* and the *intersection*. That is, given two subsets, A and B, of X there is assigned an unique subset of X according to specified rules.

In this section we shall not only define and discuss these set operations, but shall show that they possess many properties similar to the operations of arithmetic.

Definition 2.3.1

Let $A, B \subseteq X$.

a. The ***union*** of A and B, denoted by $\mathbf{A \cup B}$, is the set of all elements $x \in X$ such that x belongs to at least one of the sets A or B. Symbolically,

$$A \cup B = \{x \in X \mid (x \in A) \vee (x \in B)\}$$

b. The ***intersection*** of A and B, denoted by $\mathbf{A \cap B}$, is the set of all elements $x \in X$ such that x belongs to both A and B. Symbolically,

$$A \cap B = \{x \in X \mid (x \in A) \wedge (x \in B)\}$$

c. If $A \cap B = \emptyset$, then A and B are said to be ***disjoint.***

From the definitions of union and intersection, it should be noted that if $A, B \subseteq X$, then $A \cup B \subseteq X$ and $A \cap B \subseteq X$. Hence, if $A, B \in P(X)$, then $A \cup B, A \cap B \in P(X)$.

Example 2.3.1

a. Let X be the set of all letters of the alphabet, $A = \{a,b,c\}$ and $B = \{b,c,d\}$. Then $A, B \subseteq X$ and

$$A \cup B = \{a,b,c,d\}$$
$$A \cap B = \{b,c\}$$

b. Let $X = \mathbf{R}$ and $I = [0,1] = \{x \in \mathbf{R} \mid 0 \leq x \leq 1\}$. Then $\mathbf{N}, \mathbf{Z}, I \subseteq \mathbf{R}$ and

$$\mathbf{N} \cup \mathbf{Z} = \mathbf{Z} \qquad \mathbf{N} \cap \mathbf{Z} = \mathbf{N}$$
$$\mathbf{N} \cap I = \{1\} \qquad \mathbf{Z} \cap I = \{0,1\}$$

c. Let $X = \{1,2,3,\ldots,10\}$, $A = \{2,4,6,8\}$, $B = \{1,3,5,7,9\}$ and $C = \{3,6,9,10\}$. Then $A,B,C \subseteq X$ and

$$A \cup B = \{1,2,3,4,5,6,7,8,9\} \qquad A \cap B = \emptyset$$
$$A \cup C = \{2,3,4,6,8,9,10\} \qquad A \cap C = \{6\}$$
$$B \cup C = \{1,3,5,6,7,9,10\} \qquad B \cap C = \{3,9\}$$

Theorem 2.3.1

Let $A, B \subseteq X$. Then,

a. $A \subseteq A \cup B$ and $B \subseteq A \cup B$

b. $A \cap B \subseteq A$ and $A \cap B \subseteq B$

Proof

Let $A, B \subseteq X$.

a. We wish to show $A \subseteq A \cup B$; i.e., $(\forall x)(x \in A \Rightarrow x \in A \cup B)$. Hence, select an arbitrary element $x \in A$.

$x \in A$	
$x \in A \vee x \in B$	Addition Law
$x \in A \cup B$	Definition of Union

Therefore, $A \subseteq A \cup B$. Likewise $B \subseteq A \cup B$. ∎

b. We wish to show $A \cap B \subseteq A$; i.e., $(\forall x)(x \in A \cap B \Rightarrow x \in A)$. Hence, select an arbitrary element $x \in A \cap B$.

$x \in A \cap B$	
$x \in A \wedge x \in B$	Definition of Intersection
$x \in A$	Simplification Law

Therefore, $A \cap B \subseteq A$. Likewise, $A \cap B \subseteq B$. ∎

The above results can be represented pictorially as follows:

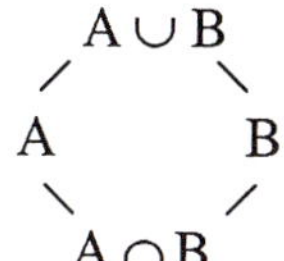

Theorem 2.3.2

Let $A, B \subseteq X$. Then,

a. $A \subseteq B \Leftrightarrow A \cup B = B$

b. $A \subseteq B \Leftrightarrow A \cap B = A$

Proof

Let $A, B \subseteq X$. We shall prove part (b) and leave the proof of part (a) as an exercise.

> **Discussion**
> Note that both part (a) and (b) involve a biconditional statement $p \Leftrightarrow q$. Recall that $(p \Leftrightarrow q) \equiv [(p \Rightarrow q) \wedge (q \Rightarrow p)]$ and that $(p \Rightarrow q) \wedge (q \Rightarrow p)$ is true only if $p \Rightarrow q$ and $q \Rightarrow p$ are both true.

b. To show $A \subseteq B \Leftrightarrow A \cap B = A$ is true, we must show the following two propositions are true:

If $A \subseteq B$, then $A \cap B = A$.
If $A \cap B = A$, then $A \subseteq B$.

($\Rightarrow$) Let $A \subseteq B$ and show $A \cap B = A$; i.e., $A \cap B \subseteq A$ and $A \subseteq A \cap B$. By Theorem 2.3.1, part (b), $A \cap B \subseteq A$. Thus it suffices to show $A \subseteq A \cap B$. Let $x \in A$.

$x \in A$	
$x \in A \Rightarrow x \in B$	$A \subseteq B$
$x \in B$	Detachment Law
$x \in A \wedge x \in B$	Definition of Conjunction
$x \in A \cap B$	Definition of Intersection

Hence, $A \subseteq A \cap B$ and thus, $A \cap B = A$.

($\Leftarrow$) Let $A \cap B = A$ and show $A \subseteq B$. Let $x \in A$.

$x \in A$	
$x \in A \cap B$	$A = A \cap B$
$x \in A \wedge x \in B$	Definition of Intersection
$x \in B$	Simplification Law

Hence, $A \subseteq B$.

Therefore, $A \subseteq B \Leftrightarrow A \cap B = A$. ■

Theorem 2.3.3

Let $A \subseteq X$. Then,

a. $A \cup \emptyset = A$

b. $A \cap \emptyset = \emptyset$

c. $A \cup X = X$

d. $A \cap X = A$

e. $A \cup A = A$

f. $A \cap A = A$

Proof

Let $A \subseteq X$. We shall use the results of Theorem 2.3.2.

a, b. Since $\emptyset \subseteq A$, $\emptyset \cup A = A$ and $\emptyset \cap A = \emptyset$. ■

c, d. Since $A \subseteq X$, $A \cup X = X$ and $A \cap X = A$. ■

e, f. Since $A \subseteq A$, $A \cup A = A$ and $A \cap A = A$. ■

Theorem 2.3.4

Let $A, B, C \subseteq X$. Then,

a. $A \cup B = B \cup A$
$A \cap B = B \cap A$ } Commutative Laws

b. $A \cup (B \cup C) = (A \cup B) \cup C$
$A \cap (B \cap C) = (A \cap B) \cap C$ } Associative Laws

c. $A \cap (B \cup C) = (A \cap B) \cup (A \cap C)$
$A \cup (B \cap C) = (A \cup B) \cap (A \cup C)$ } Distributive Laws

Proof

Let A, B, C $\subseteq$ X. The proof of part (a) is left as an exercise.

b. We shall show $A \cup (B \cup C) = (A \cup B) \cup C$ and leave the other associative law as an exercise.

We must show $A \cup (B \cup C) \subseteq (A \cup B) \cup C$ and $(A \cup B) \cup C \subseteq A \cup (B \cup C)$.

($\Rightarrow$) Let $x \in A \cup (B \cup C)$.

$x \in A \cup (B \cup C)$	$\Rightarrow (x \in A) \vee (x \in B \cup C)$	Definition of Union
	$\Rightarrow (x \in A) \vee [x \in B \vee x \in C]$	Definition of Union
	$\Rightarrow [x \in A \vee x \in B] \vee (x \in C)$	Associative Law
	$\Rightarrow (x \in A \cup B) \vee (x \in C)$	Definition of Union
	$\Rightarrow x \in (A \cup B) \cup C$	Definition of Union

Hence, $A \cup (B \cup C) \subseteq (A \cup B) \cup C$.

($\Leftarrow$) Let $x \in (A \cup B) \cup C$.

$x \in (A \cup B) \cup C$	$\Rightarrow (x \in A \cup B) \vee (x \in C)$	Definition of Union
	$\Rightarrow [x \in A \vee x \in B] \vee (x \in C)$	Definition of Union
	$\Rightarrow (x \in A) \vee [x \in B \vee x \in C]$	Associative Law
	$\Rightarrow (x \in A) \vee (x \in B \cup C)$	Definition of Union
	$\Rightarrow x \in A \cup (B \cup C)$	Definition of Union

Hence, $(A \cup B) \cup C \subseteq A \cup (B \cup C)$.

Therefore, $A \cup (B \cup C) = (A \cup B) \cup C$. ■

c. We shall show $A \cup (B \cap C) = (A \cup B) \cap (A \cup C)$ and leave the other distributive law as an exercise.

We must show $A \cup (B \cap C) \subseteq (A \cup B) \cap (A \cup C)$ and $(A \cup B) \cap (A \cup C) \subseteq A \cup (B \cap C)$.

(⇒) Let $x \in A \cup (B \cap C)$.

$$\begin{aligned} x \in A \cup (B \cap C) &\Rightarrow (x \in A) \vee [x \in (B \cap C)] \\ &\Rightarrow (x \in A) \vee [x \in B \wedge x \in C] \\ &\Rightarrow [x \in A \vee x \in B] \wedge [x \in A \vee x \in C] \\ &\Rightarrow [x \in A \cup B] \wedge [x \in A \cup C] \\ &\Rightarrow x \in (A \cup B) \cap (A \cup C) \end{aligned}$$

Hence, $A \cup (B \cap C) \subseteq (A \cup B) \cap (A \cup C)$.

(⇐) Let $x \in (A \cup B) \cap (A \cup C)$.

$$\begin{aligned} x \in (A \cup B) \cap (A \cup C) &\Rightarrow [x \in A \cup B] \wedge [x \in A \cup C] \\ &\Rightarrow [x \in A \vee x \in B] \wedge [x \in A \vee x \in C] \\ &\Rightarrow (x \in A) \vee [x \in B \wedge x \in C] \\ &\Rightarrow (x \in A) \vee [x \in B \cap C] \\ &\Rightarrow x \in A \cup (B \cap C) \end{aligned}$$

Hence, $(A \cup B) \cap (A \cup C) \subseteq A \cup (B \cap C)$.

Therefore, $A \cup (B \cap C) = (A \cup B) \cap (A \cup C)$. ■

Definition 2.3.2

Let $A, B \subseteq X$. The ***difference*** of A and B, denoted by $\mathbf{A - B}$ (read "A minus B"), is the set of all elements $x \in X$ such that x belongs to A but does not belong to B. Symbolically,

$$A - B = \{x \in X \mid (x \in A) \wedge (x \notin B)\}$$

From the definition of difference, it should be noted that if $A, B \subseteq X$, then $A - B \subseteq X$. Thus, if $A, B \in P(X)$, then $A - B \in P(X)$.

Example 2.3.2

a. Let $A = \{1, 2, \ldots, 10\}$ and $B = \{2,4,6,8,10\}$.
Then $A, B \subseteq \mathbf{N}$, $A - B = \{1,3,5,7,9\}$ and $B - A = \emptyset$ since $B \subseteq A$.

b. Let $A = \{1,2,3,4\}$ and $B = \{3,4,5\}$.
Then $A, B \subseteq \mathbf{N}$, $A - B = \{1,2\}$ and $B - A = \{5\}$.

Definition 2.3.3

Let $A \subseteq X$. The ***complement*** of A in X or the ***complement*** of A relative to X, denoted by **A′**, is the difference of X and A; i.e.,

$$A' = X - A = \{x \in X \mid x \notin A\}$$

A′ is interpreted as the set of all the elements of X that do not belong to the set A.

Example 2.3.3

a. Let $X = \{a,b,c,d,e,f,g\}$, $A = \{b,e,g\}$ and $B = \{a,b,c,d,e\}$. Then $A, B \subseteq X$ and

$A' = X - A = X - \{b,e,g\} = \{a,c,d,f\}$

$B' = X - B = X - \{a,b,c,d,e\} = \{f,g\}$

$(A \cap B)' = X - (A \cap B) = X - \{b,e\} = \{a,c,d,f,g\}$

b. Let $\mathbf{N} = \{1,2,3,4, \ldots\}$ be the set of natural numbers and $E = \{2,4,6,8, \ldots\}$ the set of even natural numbers. Then $E \subseteq \mathbf{N}$ and $E' = \mathbf{N} - E = \{1,3,5, \ldots\}$ which is the set of odd natural numbers.

Theorem 2.3.5

Let $A \subseteq X$. Then,

a. $A - \emptyset = A$

b. $\emptyset - A = \emptyset$

c. $(A')' = A$

d. $A \cup A' = X$

e. $A \cap A' = \emptyset$

Proof

Let $A \subseteq X$. We shall prove parts (a), (c) and (d) and leave the proofs of parts (b) and (e) as an exercise.

a. We must show $A - \emptyset = A$; i.e., $A - \emptyset \subseteq A$ and $A \subseteq A - \emptyset$.

(⇒) Let $x \in A - \emptyset$

$x \in A - \emptyset \Rightarrow x \in A \wedge x \notin \emptyset$	Definition of Difference
$\Rightarrow x \in A$	Simplification Law

Hence, $A - \emptyset \subseteq A$.

(⇐) Let $x \in A$

$x \in A$	
$x \notin \emptyset$	Definition of null set
$x \in A \wedge x \notin \emptyset$	Definition of Conjunction
$x \in A - \emptyset$	Definition of Difference

Hence, $A \subseteq A - \emptyset$.

Therefore, $A - \emptyset = A$. ■

c. We must show $(A')' = A$; i.e., $(A')' \subseteq A$ and $A \subseteq (A')'$.

(⇒) Let $x \in (A')'$

$x \in (A')' \Rightarrow x \notin A'$

$\Rightarrow x \in A$

Hence, $(A')' \subseteq A$.

(⇐) Let $x \in A$

$x \in A \Rightarrow x \notin A'$

$\Rightarrow x \in (A')'$

Hence, $A \subseteq (A')'$

Therefore, $(A')' = A$. ■

d. We must show $A \cup A' = X$; i.e., $A \cup A' \subseteq X$ and $X \subseteq A \cup A'$.
Since $A, A' \subseteq X$, $A \cup A' \subseteq X$. Thus it suffices to show that $X \subseteq A \cup A'$.

Let $x \in X$

$x \in X$	
$x \in A \vee x \notin A$	Tautology
$x \in A \vee x \in A'$	Definition of Complement
$x \in A \cup A'$	Definition of Union

Hence, $X \subseteq A \cup A'$.

Therefore, $A \cup A' = X$. ■

Theorem 2.3.6

Let $A, B \subseteq X$. Then,

a. $A - B = A \cap B'$

b. $A \subseteq B' \Leftrightarrow A \cap B = \emptyset$

c. $(A \cup B)' = A' \cap B'$ } DeMorgan's Laws

d. $(A \cap B)' = A' \cup B'$

Proof

Let $A, B \subseteq X$. We shall prove parts (a), (b) and (c) and leave the proof of part (d) as an exercise.

a. We must show $A - B = A \cap B'$; i.e., $A - B \subseteq A \cap B'$ and $A \cap B' \subseteq A - B$.

($\Rightarrow$) Let $x \in A - B$

$x \in A - B \Rightarrow x \in A \wedge x \notin B$	Definition of Difference
$\Rightarrow x \in A \wedge x \in B'$	Definition of Complement
$\Rightarrow x \in A \cap B'$	Definition of Intersection

Hence, $A - B \subseteq A \cap B'$

($\Leftarrow$) Let $x \in A \cap B'$

$x \in A \cap B' \Rightarrow x \in A \wedge x \in B'$	Definition of Intersection
$\Rightarrow x \in A \wedge x \notin B$	Definition of Complement
$\Rightarrow x \in A - B$	Definition of Difference

Hence $A \cap B' \subseteq A - B$.

Therefore, $A - B = A \cap B'$. ■

b. Since the statement involves the biconditional connective, we must show the following statements are true:

If $A \subseteq B'$, then $A \cap B = \emptyset$.
If $A \cap B = \emptyset$, then $A \subseteq B'$.

($\Rightarrow$) Let $A \subseteq B'$ and show $A \cap B = \emptyset$. Using an indirect method of proof, assume $A \cap B = \emptyset$ is false. Then, $\sim(A \cap B = \emptyset) \equiv (A \cap B \neq \emptyset)$ is true. Since

$A \cap B \neq \emptyset$ is assumed to be true, there exists an $x \in X$ such that $x \in A \cap B$.

$x \in A \cap B \Rightarrow x \in A \wedge x \in B$	Definition of Intersection
$\Rightarrow x \in B$	Simplification Law
$\Rightarrow x \in A$	Simplification Law
$\Rightarrow x \in B'$	$A \subseteq B'$
$\Rightarrow x \notin B$	Definition of Compliment

Hence, we have $(x \notin B) \wedge (x \in B)$ which is a contradiction. Since the contradiction arose by assuming $A \cap B = \emptyset$ was false, $A \cap B = \emptyset$ is true.

($\Leftarrow$) Let $A \cap B = \emptyset$ and show $A \subseteq B'$. Let $x \in A$.

$x \in A \Rightarrow x \notin B$	$A \cap B = \emptyset$
$\Rightarrow x \in B'$	Definition of Complement

Hence, $A \subseteq B'$.

Therefore, $A \subseteq B' \Leftrightarrow A \cap B = \emptyset$. ■

c. We must show $(A \cup B)' = A' \cap B'$; i.e., $(A \cup B)' \subseteq A' \cap B'$ and $A' \cap B' \subseteq (A \cup B)'$.

($\Rightarrow$) Let $x \in (A \cup B)'$.

$$\begin{aligned} x \in (A \cup B)' &\Rightarrow x \notin A \cup B \\ &\Rightarrow \sim (x \in A \cup B) \\ &\Rightarrow \sim (x \in A \vee x \in B) \\ &\Rightarrow \sim (x \in A) \wedge \sim (x \in B) \\ &\Rightarrow x \notin A \wedge x \notin B \\ &\Rightarrow x \in A' \wedge x \in B' \\ &\Rightarrow x \in A' \cap B' \end{aligned}$$

Hence, $(A \cup B)' \subseteq A' \cap B'$.

($\Leftarrow$) Let $x \in A' \cap B'$.

$$\begin{aligned} x \in A' \cap B' &\Rightarrow x \in A' \wedge x \in B' \\ &\Rightarrow x \notin A \wedge x \notin B \\ &\Rightarrow \sim (x \in A) \wedge \sim (x \in B) \\ &\Rightarrow \sim (x \in A \vee x \in B) \\ &\Rightarrow \sim (x \in A \cup B) \\ &\Rightarrow x \notin A \cup B \\ &\Rightarrow x \in (A \cup B)' \end{aligned}$$

Hence, $A' \cap B' \subseteq (A \cup B)'$.

Therefore, $(A \cup B)' = A' \cap B'$. ■

Example 2.3.4

Consider the power set, P(X), where $X \neq \emptyset$. The following are some properties which govern P(X) relative to $\cup$, $\cap$, $'$. For $A, B, C \in P(X)$,

$$\left.\begin{array}{l} A \cup B = B \cup A \\ A \cap B = B \cap A \end{array}\right\} \text{ Commutative Laws}$$

$$\left.\begin{array}{l} A \cup (B \cup C) = (A \cup B) \cup C \\ A \cap (B \cap C) = (A \cap B) \cap C \end{array}\right\} \text{ Associative Laws}$$

$$\left.\begin{array}{l} A \cup (B \cap C) = (A \cup B) \cap (A \cup C) \\ A \cap (B \cup C) = (A \cap B) \cup (A \cap C) \end{array}\right\} \text{ Distributive Laws}$$

$$\left.\begin{array}{l} A \cup \emptyset = A \\ A \cap X = A \end{array}\right\} \text{ Identity Laws}$$

$$\left.\begin{array}{l} A \cup A = A \\ A \cap A = A \end{array}\right\} \text{ Idempotent Laws}$$

$$\left.\begin{array}{l} A \cup A' = X \\ A \cap A' = \emptyset \end{array}\right\} \text{ Complement Laws}$$

$$\left.\begin{array}{l} (A \cup B)' = A' \cap B' \\ (A \cap B)' = A' \cup B' \end{array}\right\} \text{ DeMorgan's Laws}$$

$< P(X); \cup, \cap, ' >$ is an example of a mathematical system called an *Algebra of Sets.*

Theorem 2.3.7

Let $A, B, C \subseteq X$. Then $A \cap (B - C) = (A \cap B) - (A \cap C)$.

Proof

Rather than show $A \cap (B - C) \subseteq (A \cap B) - (A \cap C)$ and $(A \cap B) - (A \cap C) \subseteq A \cap (B - C)$, we shall make use of the results obtained from previous theorems of this section.

$(A \cap B) - (A \cap C) = (A \cap B) \cap (A \cap C)'$	Theorem 2.3.6, (a)
$= (A \cap B) \cap (A' \cup C')$	DeMorgan's Law
$= [(A \cap B) \cap A'] \cup [(A \cap B) \cap C']$	Distributive Law
$= [(A \cap A') \cap B] \cup [A \cap (B \cap C')]$	Comm. & Assoc. Laws
$= [\emptyset \cap B] \cup [A \cap (B \cap C')]$	Complement Law
$= [\emptyset] \cup [A \cap (B \cap C')]$	Theorem 2.3.3, (b)
$= A \cap (B \cap C')$	Identity Law
$= A \cap (B - C)$.	Theorem 2.3.6, (a)

■

Exercises 2.3

1. Prove part (a) of Theorem 2.3.2.

2. Prove part (a), and the remainder of part (b) and part (c) of Theorem 2.3.4.

3. Let $A, B, C, D \subseteq X$. Prove:

   a. If $A \subseteq B$ and C is any subset of X, then $A \cup C \subseteq B \cup C$.

   b. If $A \subseteq B$ and C is any subset of X, then $A \cap C \subseteq B \cap C$.

   c. If $A \subseteq B$ and $C \subseteq D$, then $A \cup C \subseteq B \cup D$.

   d. If $A \subseteq B$ and $C \subseteq D$, then $A \cap C \subseteq B \cap D$.

4. Prove part (b) and part (e) of Theorem 2.3.5.

5. Let $A, B \subseteq X$. Prove that if $A' \subseteq B$, then $A \cup B = X$.

6. Prove part (d) of Theorem 2.3.6.

7. Let $A, B \subseteq X$. Prove $A \subseteq B \Leftrightarrow B' \subseteq A'$.

In 8-10 you should make use of results obtained from previous theorems regarding the properties of $\cup$, $\cap$, $'$. (See proof of Theorem 2.3.7.)

8. Let $A, B \subseteq X$. Prove:

   a. A and $B - A$ are disjoint; i.e., $A \cap (B - A) = \emptyset$.

   b. $A \cup B = A \cup (B - A)$.

9. Let $A, B, C \subseteq X$. Prove:

   a. $A - (B \cup C) = (A - B) \cap (A - C)$.

   b. $A - (B \cap C) = (A - B) \cup (A - C)$.

10. Let $A, B \subseteq X$.
    Prove: $A - B = A - (B \cap A)$.

Section 2.4
FAMILY OF SETS/INDEXED FAMILY OF SETS

As mentioned earlier, we may think of a set as a collection of elements. Now, it may happen that the elements of a given set under consideration are themselves sets. For example, if $X = \{a,b\}$, then the power set $P(X) = \{\emptyset,\{a\},\{b\}, X\}$ is a set whose elements are themselves sets. Such sets are referred to as a *family* of subsets. In this section we shall generalize some of the concepts covered in Section 2.3 to a family of subsets.

Definition 2.4.1

Let X be a set. Then any non-empty collection of subsets of X is called a ***family*** of subsets of X.

A family of subsets of X will be denoted by script letters, such as $\mathcal{A}$, $\mathcal{B}$, ...

Example 2.4.1

a. Let $X = \{1,2,3,\ldots,10\}$. Then $\mathcal{A} = \{\{1,3,6,9\}, \{2,4,6,8\}, \{2,3,6\}, \{3,6,7\}\}$ is a family of subsets of X.

b. Let $X = \{a,b,c\}$. Then $\mathcal{B} = \{\{a\},\{b\},\{a,c\}\}$ is a family of subsets of X.

Definition 2.4.2

Let $\mathcal{A}$ be a family of subsets of X.

a. The ***union*** of $\mathcal{A}$ is defined as

$$\bigcup_{B \in \mathcal{A}} B = \{x \in X \mid \exists B \in \mathcal{A}, x \in B\}$$

b. The ***intersection*** of the sets of $\mathcal{A}$ is defined as

$$\bigcap_{B \in \mathcal{A}} B = \{x \in X \mid \forall B \in \mathcal{A}, x \in B\}$$

Note that an element $x \in X$ belongs to the union if x is an element of at least one $B \in \mathcal{A}$ while $x \in X$ belongs to the intersection only if x is an element of each $B \in \mathcal{A}$.

Example 2.4.2

Let $X = \{1,2,3, \ldots ,10\}$ and $\mathcal{A} = \{\{1,3,6,9\}, \{2,4,6,8\}, \{2,3,6\}, \{3,6,7\}\}$. Then

$$\bigcup_{B \in \mathcal{A}} B = \{1,3,6,9\} \cup \{2,4,6,8\} \cup \{2,3,6\} \cup \{3,6,7\} = \{1,2,3,4,6,7,8,9\}$$

and

$$\bigcap_{B \in \mathcal{A}} B = \{1,3,6,9\} \cap \{2,4,6,8\} \cap \{2,3,6\} \cap \{3,6,7\} = \{6\}.$$

Theorem 2.4.1

Let $\mathcal{A}$ be a family of subsets of X. Then for each $Y \in \mathcal{A}$,

a. $\bigcap_{B \in \mathcal{A}} B \subseteq Y$

b. $Y \subseteq \bigcup_{B \in \mathcal{A}} B$

Proof

This is a generalization of Theorem 2.3.1. We shall prove part (a) and leave the proof of part (b) as an exercise.

a. Let $Y \in \mathcal{A}$ and $x \in \bigcap_{B \in \mathcal{A}} B$.

$$x \in \bigcap_{B \in \mathcal{A}} B \Rightarrow (\forall B \in \mathcal{A})(x \in B)$$

$$\Rightarrow x \in Y \text{ since } Y \in \mathcal{A}.$$

Therefore, $\bigcap_{B \in \mathcal{A}} B \subseteq Y$. ■

Theorem 2.4.2

Let $\mathcal{A}$ be a family of subsets of X. Then,

a. $$\left(\bigcup_{B \in \mathcal{A}} B\right)' = \bigcap_{B \in \mathcal{A}} B'$$

b. $$\left(\bigcap_{B \in \mathcal{A}} B\right)' = \bigcup_{B \in \mathcal{A}} B'$$

Proof

This is a generalization of De Morgan's Laws. We shall prove part (b) and leave the proof of part (a) as an exercise.

b. Let $\mathcal{A}$ be a family of subsets of X. We must show

$$\left(\bigcap_{B \in \mathcal{A}} B\right)' = \bigcup_{B \in \mathcal{A}} B'; \text{ that is, } \left(\bigcap_{B \in \mathcal{A}} B\right)' \subseteq \bigcup_{B \in \mathcal{A}} B' \text{ and } \bigcup_{B \in \mathcal{A}} B' \subseteq \left(\bigcap_{B \in \mathcal{A}} B\right)'$$

($\Rightarrow$) Let $x \in \left(\bigcap_{B \in \mathcal{A}} B\right)'$.

$x \in \left(\bigcap_{B \in \mathcal{A}} B\right)' \Rightarrow x \notin \bigcap_{B \in \mathcal{A}} B$	Definition of Complement
$\Rightarrow \sim \left(x \in \bigcap_{B \in \mathcal{A}} B\right)$	Logically Equivalent
$\Rightarrow \sim [(\forall B \in \mathcal{A})(x \in B)]$	Definition of Intersection
$\Rightarrow (\exists B \in \mathcal{A})[\sim (x \in B)]$	Logically Equivalent
$\Rightarrow (\exists B \in \mathcal{A})(x \notin B)$	Logically Equivalent
$\Rightarrow (\exists B \in \mathcal{A})(x \in B')$	Definition of Complement
$\Rightarrow x \in \bigcup_{B \in \mathcal{A}} B'$	Definition of Union

Hence, $\left(\bigcap_{B \in \mathcal{A}} B\right)' \subseteq \bigcup_{B \in \mathcal{A}} B'$.

(⇐) Let $x \in \bigcup_{B \in \mathcal{A}} B'$.

$x \in \bigcup_{B \in \mathcal{A}} B' \Rightarrow (\exists B \in \mathcal{A})(x \in B')$ Definition of Union

$\Rightarrow (\exists B \in \mathcal{A})(x \notin B)$ Definition of Complement

$\Rightarrow (\exists B \in \mathcal{A})[\sim(x \in B)]$ Logically Equivalent

$\Rightarrow \sim[(\forall B \in \mathcal{A}, x \in B)]$ Logically Equivalent

$\Rightarrow \sim(x \in \bigcap_{B \in \mathcal{A}} B)$ Definition of Intersection

$\Rightarrow x \notin \bigcap_{B \in \mathcal{A}} B$ Logically Equivalent

$\Rightarrow x \in (\bigcap_{B \in \mathcal{A}} B)'$ Definition of Complement

Hence, $\bigcup_{B \in \mathcal{A}} B' \subseteq (\bigcap_{B \in \mathcal{A}} B)'$.

Therefore, $(\bigcap_{B \in \mathcal{A}} B)' = \bigcup_{B \in \mathcal{A}} B'$. ■

Example 2.4.3

Let $X = \{1,2,3,\ldots,10\}$ and $\mathcal{A} = \{\{1,3,6,9\},\{2,4,6,8\},\{2,3,6\},\{3,6,7\}\}$.

$$\begin{aligned}
\{1,3,6,9\}' &= X - \{1,3,6,9\} = \{2,4,5,7,8,10\} \\
\{2,4,6,8\}' &= X - \{2,4,6,8\} = \{1,3,5,7,9,10\} \\
\{2,3,6\}' &= X - \{2,3,6\} = \{1,4,5,7,8,9,10\} \\
\{3,6,7\}' &= X - \{3,6,7\} = \{1,2,4,5,8,9,10\}
\end{aligned}$$

Now
$$\begin{aligned}
\Big(\bigcup_{B \in \mathcal{A}} B\Big)' &= (\{1,3,6,9\} \cup \{2,4,6,8\} \cup \{2,3,6\} \cup \{3,6,7\})' \\
&= \{1,2,3,4,6,7,8,9\}' \\
&= X - \{1,2,3,4,6,7,8,9\} \\
&= \{5,10\} \\
&= \{1,3,6,9\}' \cap \{2,4,6,8\}' \cap \{2,3,6\}' \cap \{3,6,7\}' \\
&= \bigcap_{B \in \mathcal{A}} B'
\end{aligned}$$

and
$$\begin{aligned}
\Big(\bigcap_{B \in \mathcal{A}} B\Big)' &= (\{1,3,6,9\} \cap \{2,4,6,8\} \cap \{2,3,6\} \cap \{3,6,7\})' \\
&= \{6\}' \\
&= X - \{6\} \\
&= \{1,2,3,4,5,7,8,9,10\} \\
&= \{1,3,6,9\}' \cup \{2,4,6,8\}' \cup \{2,3,6\}' \cup \{3,6,7\}' \\
&= \bigcup_{B \in \mathcal{A}} B'
\end{aligned}$$

Theorem 2.4.3

Let $\mathcal{A}$ be a family of subsets of X. If $Y \subseteq X$, then,

a. $Y \cup \Big(\bigcap_{B \in \mathcal{A}} B\Big) = \bigcap_{B \in \mathcal{A}} (Y \cup B)$

b. $Y \cap \Big(\bigcup_{B \in \mathcal{A}} B\Big) = \bigcup_{B \in \mathcal{A}} (Y \cap B)$

Proof

This is a generalization of the distributive laws. We shall prove part (a) and leave the proof of part (b) as an exercise.

a. Let $\mathcal{A}$ be a family of subsets of X and $Y \subseteq X$. We must show:

$$Y \cup \left(\bigcap_{B \in \mathcal{A}} B\right) = \bigcap_{B \in \mathcal{A}} (Y \cup B); \text{ i.e.,}$$

$$Y \cup \left(\bigcap_{B \in \mathcal{A}} B\right) \subseteq \bigcap_{B \in \mathcal{A}} (Y \cup B) \text{ and } \bigcap_{B \in \mathcal{A}} (Y \cup B) \subseteq Y \cup \left(\bigcup_{B \in \mathcal{A}} B\right).$$

($\Rightarrow$) Let $x \in Y \cup \left(\bigcap_{B \in \mathcal{A}} B\right)$.

$x \in Y \cup \left(\bigcap_{B \in \mathcal{A}} B\right) \Rightarrow x \in Y \vee x \in \bigcap_{B \in \mathcal{A}} B$	Definition of Union
$\Rightarrow x \in Y \vee [(\forall B \in \mathcal{A})(x \in B)]$	Def. of Intersection
$\Rightarrow (\forall B \in \mathcal{A})(x \in Y \vee x \in B)$	Distributive Law
$\Rightarrow (\forall B \in \mathcal{A})(x \in Y \cup B)$	Definition of Union
$\Rightarrow x \in \bigcap_{B \in \mathcal{A}} (Y \cup B)$	Def. of Intersection

Hence, $Y \cup \left(\bigcap_{B \in \mathcal{A}} B\right) \subseteq \bigcap_{B \in \mathcal{A}} (Y \cup B)$.

($\Leftarrow$) Let $x \in \bigcap_{B \in \mathcal{A}} (Y \cup B)$.

$x \in \bigcap_{B \in \mathcal{A}} (Y \cup B) \Rightarrow (\forall B \in \mathcal{A})(x \in Y \cup B)$	Def. of Intersection
$\Rightarrow (\forall B \in \mathcal{A})(x \in Y \vee x \in B)$	Definition of Union
$\Rightarrow x \in Y \vee [(\forall B \in \mathcal{A})(x \in B)]$	Distributive Law
$\Rightarrow x \in Y \vee x \in \bigcap_{B \in \mathcal{A}} B$	Def. of Intersection
$\Rightarrow x \in Y \cup \left(\bigcap_{B \in \mathcal{A}} B\right)$	Definition of Union

Hence, $\bigcap_{B \in \mathcal{A}} (Y \cup B) \subseteq Y \cup \left(\bigcap_{B \in \mathcal{A}} B\right)$.

Therefore, $Y \cup \left(\bigcap_{B \in \mathcal{A}} B\right) = \bigcap_{B \in \mathcal{A}} (Y \cup B)$. ■

Given a family $\mathcal{A}$ of subsets of X, it is often possible to associate some sort of "index" with each element of the family. In this way each element or set of the family can be identified by its index. For example, let X = {a,b,c} and $\mathcal{A}$ = {{a},{b},{a,c}} be a family of subsets of X. If we let B_1 = {a}, B_2 = {b} and B_3 = {a,c}, then $\mathcal{A} = \{B_1,B_2,B_3\}$ and each set of $\mathcal{A}$ is identified by its index 1, 2 or 3.

Definition 2.4.3

Let **I** be a non-empty set. Suppose for each $i \in \mathbf{I}$, there corresponds a subset A_i of X. Then the set

$$\mathcal{A} = \{A_i \mid i \in \mathbf{I}\}$$

is called an ***indexed*** family of subsets of X, **I** the ***indexing set*** and each $i \in \mathbf{I}$ an ***index***.

Example 2.4.4

a. Let $\mathcal{A} = \{\mathbf{N},\mathbf{Z},\mathbf{Q},\mathbf{R}\}$ be a family of subsets of **R** and $A_1 = \mathbf{N}$, $A_2 = \mathbf{Z}$, $A_3 = \mathbf{Q}$, and $A_4 = \mathbf{R}$. Then

$$\mathcal{A} = \{A_1,A_2,A_3,A_4\} = \{A_i \mid i \in \mathbf{I}\}$$

is an indexed family of subsets of **R** indexed by the set **I** = {1,2,3,4}.

b. For each $i \in \mathbf{N}$, consider the open interval $A_i = (0,i) = \{x \in \mathbf{R} \mid 0 < x < i\} \subseteq \mathbf{R}$. Then

$$\mathcal{A} = \{A_1,A_2,A_3,\ldots\} = \{A_i \mid i \in \mathbf{N}\}$$

is an indexed family of subsets of **R** indexed by the set **N** = {1,2,3, . . .}. The set $A_2 = (0,2)$ has index 2 while $A_{40} = (0,40)$ has index 40.

c. For each $i \in \mathbf{I} = \{-1,0,1,2,3,\ldots\}$, consider the set $A_i = \{i,i+1\} \subseteq \mathbf{Z}$. Then

$$\mathcal{A} = \{A_{-1},A_0,A_1,\ldots\} = \{A_i \mid i \in \mathbf{I}\}$$

is an indexed family of subsets of **Z** indexed by the set **I**. The set $A_{-1} = \{-1,0\}$ has index -1 while $A_{15} = \{15,16\}$ has index 15.

Definition 2.4.4

Let $\mathcal{A} = \{A_i \mid i \in I\}$ be an indexed family of subsets of X. Then,

a. The ***union*** of $\mathcal{A}$ is defined as

$$\bigcup_{i \in \mathbf{I}} A_i = \{x \in X \mid \exists i \in \mathbf{I}, x \in A_i\}.$$

b. The ***intersection*** of the sets of $\mathcal{A}$ is defined as

$$\bigcap_{i \in \mathbf{I}} A_i = \{x \in X \mid \forall i \in \mathbf{I}, x \in A_i\}.$$

Example 2.4.5

Refer to Example 2.4.4.

a. In part (a) we have:

$$\bigcup_{i \in \mathbf{I}} A_i = A_1 \cup A_2 \cup A_3 \cup A_4$$
$$= \mathbf{N} \cup \mathbf{Z} \cup \mathbf{Q} \cup \mathbf{R}$$
$$= \mathbf{R} \text{ since } \mathbf{N} \subset \mathbf{Z} \subset \mathbf{Q} \subset \mathbf{R}.$$

and

$$\bigcap_{i \in \mathbf{I}} A_i = A_1 \cap A_2 \cap A_3 \cap A_4$$
$$= \mathbf{N} \cap \mathbf{Z} \cap \mathbf{Q} \cap \mathbf{R}$$
$$= \mathbf{N} \text{ since } \mathbf{N} \subset \mathbf{Z} \subset \mathbf{Q} \subset \mathbf{R}.$$

b. In part (b), since $A_i \subset A_{i+1}$, we have:

$$\bigcup_{i \in \mathbf{N}} A_i = A_1 \cup A_2 \cup A_3 \cup \ldots = (0,1) \cup (0,2) \cup (0,3) \cup \ldots = (0,\infty)$$

$$\bigcap_{i \in \mathbf{N}} A_i = A_1 \cap A_2 \cap A_3 \cap \ldots = (0,1) \cap (0,2) \cap (0,3) \cap \ldots = (0,1)$$

c. In part (c) we have:

$$\bigcup_{i \in \mathbf{I}} A_i = A_{-1} \cup A_0 \cup A_1 \cup \ldots = \{-1,0\} \cup \{0,1\} \cup \{1,2\} \cup \ldots$$
$$= \{-1,0,1,2 \ldots\} = \mathbf{I}$$

$$\bigcap_{i \in \mathbf{I}} A_i = A_{-1} \cap A_0 \cap A_1 \cap \ldots = \{-1,0\} \cap \{0,1\} \cap \{1,2\} \cap \ldots = \emptyset$$

The following theorem is a direct consequence of Theorems 2.4.1, 2.4.2, and 2.4.3.

Theorem 2.4.4

Let $\mathcal{A} = \{A_i \mid i \in \mathbf{I}\}$ be an indexed family of subsets of X and $Y \subseteq X$. Then

a. $\bigcap_{i \in \mathbf{I}} A_i \subseteq A_i$ and $A_i \subseteq \bigcup_{i \in \mathbf{I}} A_i$, $\forall i \in \mathbf{I}$

b. $(\bigcup_{i \in \mathbf{I}} A_i)' = \bigcap_{i \in \mathbf{I}} A_i{}'$ and $(\bigcap_{i \in \mathbf{I}} A_i)' = \bigcup_{i \in \mathbf{I}} A_i{}'$

c. $Y \cup (\bigcap_{i \in \mathbf{I}} A_i) = \bigcap_{i \in \mathbf{I}} (Y \cup A_i)$ and $Y \cap (\bigcup_{i \in \mathbf{I}} A_i) = \bigcup_{i \in \mathbf{I}} (Y \cap A_i)$

Definition 2.4.5

Let $\mathcal{A} = \{A_i \mid i \in \mathbf{I}\}$ be an indexed family of subsets of X.

a. If $n \in \mathbf{N}$ and if $\mathbf{I} = \{1, 2, \ldots, n\}$ is the indexing set of $\mathcal{A}$; i.e., $\mathcal{A} = \{A_1, A_2, \ldots, A_n\}$, we sometimes write

$$A_1 \cup A_2 \cup \ldots \cup A_n \text{ or } \bigcup_{i=1}^{n} A_i \text{ for } \bigcup_{i \in \mathbf{I}} A_i$$

and

$$A_1 \cap A_2 \cap \ldots \cap A_n \text{ or } \bigcap_{i=1}^{n} A_i \text{ for } \bigcap_{i \in \mathbf{I}} A_i.$$

b. If $k, n \in \mathbf{N}$, $k < n$, and if $\mathbf{I} = \{k, k+1, \ldots, n\}$ is the indexing set of $\mathcal{A}$; i.e., $\mathcal{A} = \{A_k, A_{k+1}, \ldots, A_n\}$, we sometimes write

$$A_k \cup A_{k+1} \cup \ldots \cup A_n \text{ or } \bigcup_{i=k}^{n} A_i \text{ for } \bigcup_{i \in \mathbf{I}} A_i$$

and

$$A_k \cap A_{k+1} \cap \ldots \cap A_n \text{ or } \bigcap_{i=k}^{n} A_i \text{ for } \bigcap_{i \in \mathbf{I}} A_i.$$

Example 2.4.6

Let $X = \{1,2,\ldots,10,11\}$, $A_i = \{1,2,\ldots,i\}$ and $\mathcal{A} = \{A_i \mid i = 1,2,\ldots,10\}$. Then for each i, $i = 1,2,\ldots,10$, $A_i \subseteq X$ and $\mathcal{A} = \{A_1,A_2,\ldots,A_{10}\}$.

a. Now
$$\begin{aligned} A_1 &= \{1\} \\ A_2 &= \{1,2\} \\ A_3 &= \{1,2,3\} \\ &\vdots \\ A_{10} &= \{1,2,3,\ldots,10\}. \end{aligned}$$

Thus we have

$$\bigcup_{i=1}^{10} A_i = A_1 \cup A_2 \cup A_3 \cup \ldots \cup A_{10} = \{1,2,3,\ldots,10\}$$

$$\bigcap_{i=1}^{10} A_i = A_1 \cap A_2 \cap A_3 \cap \ldots \cap A_{10} = \{1\}.$$

b. Now
$$\begin{aligned} A_1' &= \{1\}' &&= \{2,3,\ldots,10,11\} \\ A_2' &= \{1,2\}' &&= \{3,4,\ldots,10,11\} \\ A_3' &= \{1,2,3\}' &&= \{4,5,\ldots,10,11\} \\ &\vdots && \vdots \\ A_{10}' &= \{1,2,\ldots,10\}' = \{11\}. \end{aligned}$$

Applying the results of Theorem 2.4.4, part (b), we have

$$\left(\bigcup_{i=1}^{10} A_i\right)' = (\{1,2,3,\ldots,10\})' = \{11\} = \bigcap_{i=1}^{10} A_i'$$

and

$$\left(\bigcap_{i=1}^{10} A_i\right)' = (\{1\})' = \{2,3,\ldots,10,11\} = \bigcup_{i=1}^{10} A_i'$$

Example 2.4.6 continued next page.

Example 2.4.6 Continued

c. If $Y = \{2,4,6\} \subseteq X$, then

$$\begin{array}{ll} Y \cup A_1 = \{1,2,4,6\} & Y \cap A_1 = \emptyset \\ Y \cup A_2 = \{1,2,4,6\} & Y \cap A_2 = \{2\} \\ Y \cup A_3 = \{1,2,3,4,6\} & Y \cap A_3 = \{2\} \\ \vdots & \vdots \\ Y \cup A_{10} = \{1,2,3,\ldots,10\} & Y \cap A_{10} = \{2,4,6\} \end{array}$$

Applying the results of Theorem 2.4.4, part (c), we have

$$Y \cup \left(\bigcap_{i=1}^{10} A_i \right) = \{2,4,6\} \cup \{1\} = \{1,2,4,6\} = \bigcap_{i=1}^{10} (Y \cup A_i)$$

$$Y \cap \left(\bigcup_{i=1}^{10} A_i \right) = \{2,4,6\} \cap \{1,2,3,\ldots,10\} = \{2,4,6\} = \bigcup_{i=1}^{10} (Y \cap A_i)$$

Definition 2.4.6

Let $\mathcal{A} = \{A_i \mid i \in \mathbf{I}\}$ be an indexed family of subsets of X. The family is said to be ***pairwise disjoint*** if $\forall i, j \in \mathbf{I}$, whenever $A_i \neq A_j$, then $A_i \cap A_j = \emptyset$.

Let $\mathcal{A} = \{A_i \mid i \in \mathbf{I}\}$ be any indexed family of subsets of X. If $\mathcal{A}$ is pairwise disjoint, then, by definition, when $A_i \neq A_j$, $A_i \cap A_j = \emptyset$. Recall that $(p \Rightarrow q) \equiv (\sim p \vee q)$. Therefore,

$$\begin{aligned} [(A_i \neq A_j) \Rightarrow (A_i \cap A_j = \emptyset)] &\equiv [\sim(A_i \neq A_j) \vee (A_i \cap A_j = \emptyset)] \\ &\equiv [(A_i = A_j) \vee (A_i \cap A_j = \emptyset)] \end{aligned}$$

Hence, $\mathcal{A}$ is pairwise disjoint if $\forall i, j \in \mathbf{I}$ either $A_i = A_j$ or $A_i \cap A_j = \emptyset$.

Example 2.4.7

Let $X = \{1,2,3,\ldots,10\}$.

a. If $\mathcal{A} = \{A_1,A_2,A_3\}$ where $A_1 = \{1,2\}$, $A_2 = \{2,3\}$ and $A_3 = \{4,5\}$, then the family of subsets, $\mathcal{A}$, is not pairwise disjoint since $A_1 \neq A_2$ but $A_1 \cap A_2 = \{2\} \neq \emptyset$.

b. If $\mathcal{B} = \{B_1,B_2,B_3\}$ where $B_1 = \{1,2\}$, $B_2 = \{3,4\}$, and $B_3 = \{5,6\}$, then the family of subsets, $\mathcal{B}$, is pairwise disjoint since $B_1 \neq B_2$, $B_1 \neq B_3$, $B_2 \neq B_3$ and $B_1 \cap B_2 = B_1 \cap B_3 = B_2 \cap B_3 = \emptyset$.

Let $\mathcal{A} = \{A_i \mid i \in \mathbf{I}\}$ be a family of subsets of X having at least two distinct elements (sets). If the family is pairwise disjoint, then $\bigcap_{i \in \mathbf{I}} A_i = \emptyset$. However, if $\bigcap_{i \in \mathbf{I}} A_i = \emptyset$, the family need not be pairwise disjoint. (See Example 2.4.7, part (a), above.)

Exercises 2.4

1. Prove part (b) of Theorem 2.4.1.

2. Let $X = \{1,2,3,\ldots,15\}$ and $\mathcal{A}$ be a family of subsets of X where:

   $\mathcal{A} = \{\{4,6,8,10,12\}, \{1,4,8,12\}, \{1,3,4,5,7,11,12\}, \{1,2,4,7,12,15\}\}$.

   Find:

   a. $\bigcup_{B \in \mathcal{A}} B$ b. $\bigcap_{B \in \mathcal{A}} B$

   c. $\bigcap_{B \in \mathcal{A}} B'$

3. Let $\mathcal{A}$ be a family of subsets of X. Prove:

   a. $\bigcup_{B \in \mathcal{A}} B \subseteq X$ b. $\bigcap_{B \in \mathcal{A}} B \subseteq X$

4. Prove part (a) of Theorem 2.4.2.

5. Prove part (b) of Theorem 2.4.3.

6. For each $k \in \mathbf{N}$, consider the closed interval

   $A_k = [0,\frac{1}{k}] = \{x \in \mathbf{R} \mid 0 \leq x \leq \frac{1}{k}\}$

   Determine:

   a. $\bigcap_{k=1}^{100} A_k$ b. $\bigcup_{k=1}^{100} A_k$

   c. $\bigcap_{k \in \mathbf{N}} A_k$ d. $\bigcup_{k \in \mathbf{N}} A_k$

Section 2.5
PRODUCT SETS

Let a and b be two elements. Suppose we wish to impose an order on these two elements by designating one of the elements, say a, as the first element and the other element, b, as the second element. Such a designation will be referred to as an *ordered pair* and denoted symbolically by **(a,b)**.

Definition 2.5.1

Let A and B be two sets. The ***product set*** of A and B, denoted by **A x B** (read "A cross B"), is the set of all ordered pairs (a,b) where $a \in A$ and $b \in B$. Symbolically,

$$A \times B = \{(a,b) \mid a \in A \wedge b \in B\}$$

Definition 2.5.2

Let A and B be two sets and $(a,b), (c,d) \in A \times B$. Then (a,b) and (c,d) are said to be ***equal*** and we write **(a,b) = (c,d)** if a = c and b = d.

$$(a,b) = (c,d) \Leftrightarrow a = c \text{ and } b = d$$

Note that if $(a, b) \neq (c, d)$, then the statement $(a = c) \wedge (b = d)$ is false. Hence, $\sim [(a = c) \wedge (b = d)]$ is true. But,

$$\sim [(a = c) \wedge (b = d)] \equiv \sim (a = c) \vee \sim (b = d)$$

$$\equiv (a \neq c) \vee (b \neq d).$$

Thus, to show $(a,b) \neq (c,d)$, one must show that either $a \neq c$ or $b \neq d$.

Example 2.5.1

a. Let $A = \{a,b,c\}$ and $B = \{d,e\}$.

$A \times B = \{(a,d), (a,e), (b,d), (b,e), (c,d), (c,e)\}$.

$B \times A = \{(d,a), (d,b), (d,c), (e,a), (e,b), (e,c)\}$.

Hence, $A \times B \neq B \times A$.

b. Let $A = \{1,2,3\}$ and $B = \emptyset$. Since $B = \emptyset$, $A \times B = \emptyset$ for there does not exist an ordered pair (a,b) where $a \in A$ and $b \in B$.

Theorem 2.5.1

Let A, B, C and D be sets.

a. If $A \neq \emptyset$ and $B \neq \emptyset$, then $A \times B \neq \emptyset$.

b. If $A \subseteq C$ and $B \subseteq D$, then $A \times B \subseteq C \times D$.

Proof

The proof of part (a) is obvious. Therefore, we shall prove part (b).

b. Let $A \subseteq C$ and $B \subseteq D$. We must show $A \times B \subseteq C \times D$.

Let $(x, y) \in A \times B$.

$(x, y) \in A \times B$	$\Rightarrow x \in A \wedge y \in B$	Definition
	$\Rightarrow x \in A$	Simplification Law
	$\Rightarrow y \in B$	Simplification Law
	$\Rightarrow x \in C$	$A \subseteq C$
	$\Rightarrow y \in D$	$B \subseteq D$
	$\Rightarrow x \in C \wedge y \in D$	Conjunction
	$\Rightarrow (x,y) \in C \times D$	Definition

Therefore, $A \times B \subseteq C \times D$. ■

Theorem 2.5.2

Let A, B, C $\subseteq$ X. Then,

a. $A \times (B \cup C) = (A \times B) \cup (A \times C)$

b. $A \times (B \cap C) = (A \times B) \cap (A \times C)$

c. $A \times (B - C) = (A \times B) - (A \times C)$

Proof

Let A, B, C $\subseteq$ X. We shall prove parts (b) and (c), and leave the proof of part (a) as an exercise.

b. We must show that $A \times (B \cap C) = (A \times B) \cap (A \times C)$.

($\Rightarrow$) Let $(x,y) \in A \times (B \cap C)$.

$$
\begin{aligned}
(x,y) \in A \times (B \cap C) &\Rightarrow x \in A \wedge y \in (B \cap C) \\
&\Rightarrow x \in A \wedge (y \in B \wedge y \in C) \\
&\Rightarrow (x \in A \wedge x \in A) \wedge (y \in B \wedge y \in C) \\
&\Rightarrow (x \in A \wedge y \in B) \wedge (x \in A \wedge y \in C) \\
&\Rightarrow (x,y) \in A \times B \wedge (x, y) \in A \times C \\
&\Rightarrow (x,y) \in (A \times B) \cap (A \times C)
\end{aligned}
$$

Hence, $A \times (B \cap C) \subseteq (A \times B) \cap (A \times C)$.

($\Leftarrow$) Let $(x, y) \in (A \times B) \cap (A \times C)$.

$$
\begin{aligned}
(x, y) \in (A \times B) \cap (A \times C) &\Rightarrow (x, y) \in A \times B \wedge (x, y) \in A \times C \\
&\Rightarrow (x \in A \wedge y \in B) \wedge (x \in A \wedge y \in C) \\
&\Rightarrow (x \in A \wedge x \in A) \wedge (y \in B \wedge y \in C) \\
&\Rightarrow x \in A \wedge y \in B \cap C \\
&\Rightarrow (x,y) \in A \times (B \cap C)
\end{aligned}
$$

Hence, $(A \times B) \cap (A \times C) \subseteq A \times (B \cap C)$.

Therefore, $A \times (B \cap C) = (A \times B) \cap (A \times C)$. ■

c. We must show $A \times (B - C) = (A \times B) - (A \times C)$.

($\Rightarrow$) Let $(x, y) \in A \times (B - C)$.

$$
\begin{aligned}
(x, y) \in A \times (B - C) &\Rightarrow x \in A \wedge y \in (B - C) \\
&\Rightarrow x \in A \wedge (y \in B \wedge y \notin C) \\
&\Rightarrow (x \in A \wedge x \in A) \wedge (y \in B \wedge y \notin C) \\
&\Rightarrow (x \in A \wedge y \in B) \wedge (x \in A \wedge y \notin C) \\
&\Rightarrow (x,y) \in (A \times B) \wedge (x,y) \notin (A \times C) \\
&\Rightarrow (x,y) \in (A \times B) - (A \times C)
\end{aligned}
$$

Hence, $A \times (B - C) \subseteq (A \times B) - (A \times C)$.

($\Leftarrow$) Let $(x, y) \in (A \times B) - (A \times C)$.

$$\begin{aligned}(x, y) \in (A \times B) - (A \times C) &\Rightarrow (x, y) \in A \times B \wedge (x, y) \notin A \times C \\ &\Rightarrow (x \in A \wedge y \in B) \wedge (x \in A \wedge y \notin C) \\ &\Rightarrow (x \in A \wedge x \in A) \wedge (y \in B \wedge y \notin C) \\ &\Rightarrow x \in A \wedge y \in B - C \\ &\Rightarrow (x,y) \in A \times (B - C)\end{aligned}$$

Hence, $(A \times B) - (A \times C) \subseteq A \times (B - C)$.

Therefore, $A \times (B - C) = (A \times B) - (A \times C)$. ■

Example 2.5.2

The following illustrates properties (b) and (c) of Theorem 2.5.2.
Let $X = \{1,2, \ldots, 10\}$, $A = \{6,7\}$, $B = \{1,2,3\}$ and $C = \{2,4\}$.

a. $A \times (B \cap C) = (A \times B) \cap (A \times C)$

$$\begin{aligned}A \times (B \cap C) &= \{6,7\} \times (\{1,2,3\} \cap \{2,4\}) \\ &= \{6,7\} \times \{2\} \\ &= \{(6,2), (7,2)\}\end{aligned}$$

$(A \times B) \cap (A \times C) = (\{6,7\} \times \{1,2,3\}) \cap (\{6,7\} \times \{2,4\})$
$= \{(6,1),(6,2),(6,3),(7,1),(7,2),(7,3)\} \cap \{(6,2),(6,4),(7,2),(7,4)\}$
$= \{(6,2), (7,2)\}$

Therefore, $A \times (B \cap C) = (A \times B) \cap (A \times C)$.

b. $A \times (B - C) = (A \times B) - (A \times C)$

$$\begin{aligned}A \times (B - C) &= \{6,7\} \times (\{1,2,3\} - \{2,4\}) \\ &= \{6,7\} \times \{1,3\} \\ &= \{(6,1), (6,3), (7,1), (7,3)\}\end{aligned}$$

$(A \times B) - (A \times C) = (\{6,7\} \times \{1,2,3\}) - (\{6,7\} \times \{2,4\})$
$= \{(6,1),(6,2),(6,3),(7,1),(7,2),(7,3)\} - \{(6,2),(6,4),(7,2),(7,4)\}$
$= \{(6,1), (6,3), (7,1), (7,3)\}$

Therefore, $A \times (B - C) = (A \times B) - (A \times C)$.

Theorem 2.5.3

Let $A \subseteq X$ and $\mathcal{B} = \{B_i \mid i \in \mathbf{I}\}$ be an indexed family of subsets of X. Then,

a. $A \times (\bigcup_{i \in \mathbf{I}} B_i) = \bigcup_{i \in \mathbf{I}} (A \times B_i)$

b. $A \times (\bigcap_{i \in \mathbf{I}} B_i) = \bigcap_{i \in \mathbf{I}} (A \times B_i)$

Proof

Note that this theorem is an extension of Theorem 2.5.2, parts (a) and (b). We shall prove part (a) and leave the proof of part (b) as an exercise.

a. Let $A \subseteq X$ and $\mathcal{B} = \{B_i \mid i \in \mathbf{I}\}$ be an indexed family of subsets of X. We must show:

$$A \times (\bigcup_{i \in \mathbf{I}} B_i) = \bigcup_{i \in \mathbf{I}} (A \times B_i).$$

($\Rightarrow$) Let $(x,y) \in A \times (\bigcup_{i \in \mathbf{I}} B_i)$.

$$\begin{aligned}(x,y) \in A \times (\bigcup_{i \in \mathbf{I}} B_i) &\Rightarrow x \in A \wedge y \in (\bigcup_{i \in \mathbf{I}} B_i)\\ &\Rightarrow (x \in A) \wedge (\exists i \in \mathbf{I})(y \in B_i)\\ &\Rightarrow (\exists i \in \mathbf{I})(x \in A \wedge y \in B_i)\\ &\Rightarrow (\exists i \in \mathbf{I})[(x,y) \in A \times B_i]\\ &\Rightarrow (x,y) \in \bigcup_{i \in \mathbf{I}} (A \times B_i)\end{aligned}$$

Hence, $A \times (\bigcup_{i \in \mathbf{I}} B_i) \subseteq \bigcup_{i \in \mathbf{I}} (A \times B_i)$.

($\Leftarrow$) Let $(x,y) \in \bigcup_{i \in \mathbf{I}} (A \times B_i)$.

$$\begin{aligned}(x,y) \in \bigcup_{i \in \mathbf{I}} (A \times B_i) &\Rightarrow (\exists i \in \mathbf{I})[(x,y) \in A \times B_i] \\ &\Rightarrow (\exists i \in \mathbf{I})(x \in A \wedge y \in B_i) \\ &\Rightarrow (x \in A) \wedge [(\exists i \in \mathbf{I})(y \in B_i)] \\ &\Rightarrow x \in A \wedge y \in \bigcup_{i \in \mathbf{I}} B_i \\ &\Rightarrow (x,y) \in A \times (\bigcup_{i \in \mathbf{I}} B_i)\end{aligned}$$

Hence, $\bigcup_{i \in \mathbf{I}} (A \times B_i) \subseteq A \times (\bigcup_{i \in \mathbf{I}} B_i)$.

Therefore, $A \times (\bigcup_{i \in \mathbf{I}} B_i) = \bigcup_{i \in \mathbf{I}} (A \times B_i)$. ■

Exercises 2.5

1. Prove part (a) of Theorem 2.5.2.

2. Prove part (b) of Theorem 2.5.3.

3. Let $A = \{1,2\}$, $B = \{4,5,6\}$, $C = \{1,2,3\}$ and $D = \{3,4,5,6\}$; where $A \subseteq C$ and $B \subseteq D$.

   Verify: $A \times B \subseteq C \times D$.

4. Using the sets: X, A, B and C as defined in Example 2.5.2, verify the following:

   $A \times (B \cup C) = (A \times B) \cup (A \times C)$.

5. Let $A_1, A_2, \ldots, A_n$ be n sets. The *product set* of $A_1, A_2, \ldots, A_n$, denoted by $A_1 \times A_2 \times \ldots \times A_n$, is the set of all ordered n-tuples $(a_1, a_2, \ldots, a_n)$ where $a_i \in A_i$. Symbolically,

   $A_1 \times A_2 \times \ldots \times A_n$
   $= \{(a_1, a_2, \ldots, a_n) \mid a_i \in A_i\}$.

   Let $H = \{p,q\}$, $K = \{r,s\}$ and $L = \{t,u,v\}$

   a. List of elements of $H \times K \times L$.

   b. List the elements of the set $H \times (K \times L)$.

   c. List the elements of the set $(H \times K) \times L$.

   d. Discuss whether the sets $H \times K \times L$, $H \times (K \times L)$ and $(H \times K) \times L$ are the same. What can be said about the number of elements in each set?

Chapter 3

FUNCTIONS

The concept of a function is probably one of the most important in the field of mathematics. It is utilized in all branches of mathematics to specify a special type of correspondence between sets. It is the basis for relating two sets in order to investigate and compare their properties. In this chapter we introduce and discuss a function together with its main properties. Our approach to the concept of a function will be in the manner in which it evolved. Leonhard Euler introduced the notation of a function, which is now standard, in 1750 and Lejeune Dirichlet refined the definition of a function in 1837. Dirichlet's refined definition is the basis of our definition of a function.

Georg Cantor introduced the definition of a function in terms of ordered pairs of elements over fifty years later and we shall discuss this alternate approach in Chapter 4.

Section 3.1
BASIC CONCEPTS

From your previous studies in mathematics you have probably developed an intuitive sense for a function. You may be familiar with formulas (algebraic equations) involving two variables, say x and y, where x and y represented real numbers. In that context, values of a variable y were determined by values of a variable x using some definite rule; such as $y = 2x - 1$ or $y = x^2$ or $y = \sqrt{2x - 1}$. We shall generalize this idea to formalize the definition of a function for any two arbitrary sets of elements.

Definition 3.1.1

Let A and B be non-empty sets. A ***function*** from set A to set B is a correspondence that assigns to each element $a \in A$ an unique (one and only one) element $b \in B$.

Letters such as f, g, h, etc., or F, G, H, etc. will be used to denote functions. We shall write **f: A → B** to indicate f is a function from A to B.

Let $f: A \to B$. If f assigns to the element $a \in A$ the element $b \in B$, then b is called the *image* of a under f and we write $b = f(a)$ (read "b equals f of a"). This idea is pictured in the following figure.

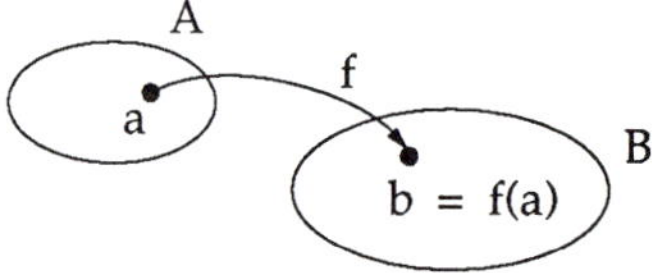

By the definition of a function, every element $a \in A$ has an uniquely determined $b \in B$ such that $f(a) = b$. Specifically, if $f(a) = b$ and $f(a) = c$, then $b = c$.

Example 3.1.1

Let A = {a,b,c} and B = {1,2,3}

a. Consider the following:

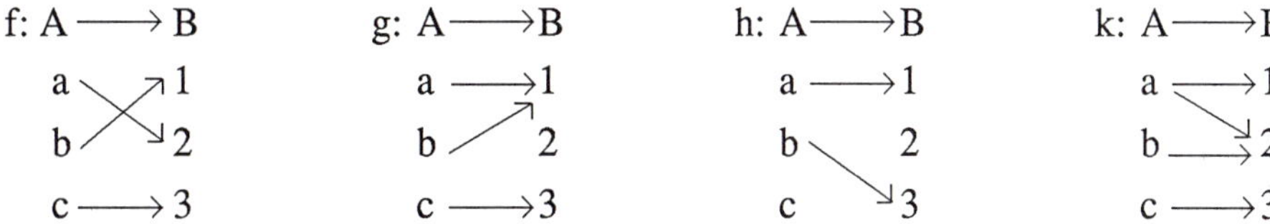

We see that f and g are functions from A to B. However h is not a function from A to B, since h(c) is not defined (h has not assigned c to any element of B). Also, k is not a function from A to B since k(a) = 1 and k(a) = 2, but $1 \neq 2$ (k has not assigned a to a unique element of B).

b. Since the correspondences f and g in part (a) are functions from A to B, the images of the elements in A under both f and g may be written in the form f(x) or g(x) where $x \in A$.

f: A ⟶ B
a ↗ 1 = f (b)
b ↘ 2 = f (a)
c ⟶ 3 = f (c)

g: A ⟶ B
a ⟶ 1 = g(a) = g(b)
b ↗ 2
c ⟶ 3 = g (c)

Definition 3.1.2

Let f: A → B.

a. The set A is called the ***domain*** of f and is denoted by $\mathbf{D_f}$.

b. The ***range*** of f, denoted by $\mathbf{R_f}$, is the subset of B which contains all the images of the elements of A under f. Symbolically,

$$R_f = \{f(a) \in B \mid a \in A\}.$$

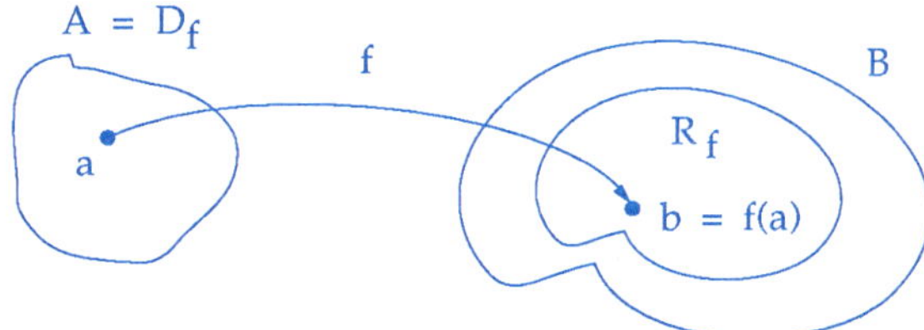

Note that if $b \in R_f$, then there must exist at least one $a \in A$ such that b = f(a).

Example 3.1.2

a. Refer to Example 3.1.1 where A = {a,b,c}, B = {1,2,3} and

f: A ⟶ B

a ↘ 2, b ↗ 1, c ⟶ 3

g: A ⟶ B

a ⟶ 1, b ↗ 1, c ⟶ 3

Now $D_f = A$ and $R_f = \{f(a),f(b),f(c)\} = \{2,1,3\} = B$ while $D_g = A$ and $R_g = \{g(a),g(b),g(c)\} = \{1,3\} \subset B$.

b. Let f: **R** ⟶ **R** where $f(x) = x^2$. Then $D_f = \mathbf{R}$ and $R_f = [0,\infty) \subset \mathbf{R}$.

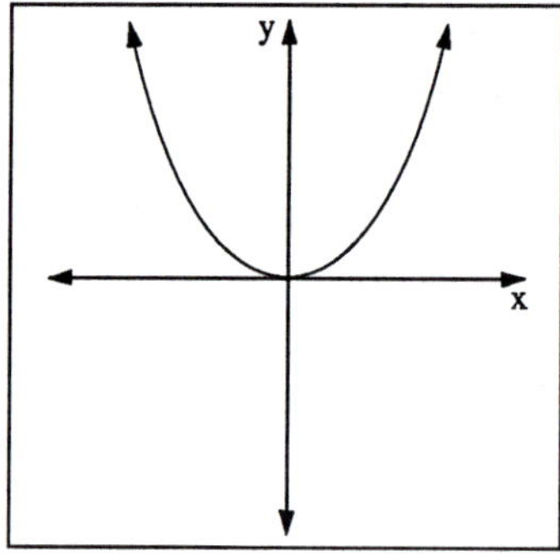

c. Let g: **R** – {0} ⟶ **R** where $g(x) = \frac{1}{x}$.
Then $D_g = \mathbf{R} - \{0\}$ and $R_g = \mathbf{R} - \{0\} \subset \mathbf{R}$.

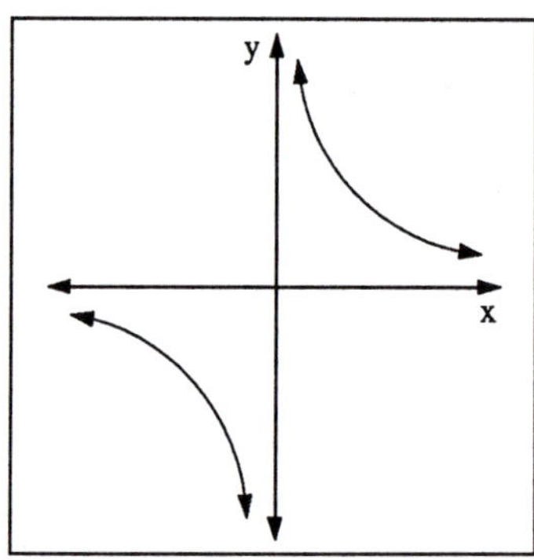

Definition 3.1.3

a. Let $A \subseteq X$, $X \neq \emptyset$ and $\mathbf{X}_A$: $X \to \{0,1\}$ where for each $x \in X$,

$$\mathbf{X}_A(x) = \begin{cases} 0 & \text{when } x \notin A \\ 1 & \text{when } x \in A \end{cases}$$

Then $\mathbf{X}_A$ is called the ***characteristic*** function of A in X.

b. Let **e**: A ⟶ A where **e**(a) = a, ∀a ∈ A. Then **e** is called the ***identity*** function on A.

c. Let A and B be two non-empty sets and b a fixed element of B. Consider the function f: A ⟶ B where f(a) = b, ∀a ∈ A. Then f is called a ***constant*** function.

Example 3.1.3

a. Let X = {1,2,3,4,5} and A = {2,4}. Then the function $\mathbf{X}_A$: X ⟶ {0,1} where,

$\mathbf{X}_A$: X ⟶ {0,1}

1 ⟶ 0
2 ⟶ 1
3 ⟶ 0
4 ⟶ 1
5 ⟶ 0

is the *characteristic* function of A in X.

b. Let A = {1,2,3}. Then the function **e**: A ⟶ A where,

e: A ⟶ A
1 ⟶ 1 = **e** (1)
2 ⟶ 2 = **e** (2)
3 ⟶ 3 = **e** (3)

is the *identity* function on A since ∀a ∈ A, **e**(a) = a.

c. Let A = {a,b,c} and B = {1,2,3}. Then the function f: A ⟶ B where,

f: A ⟶ B
a ⟶ 1
b ⟶ 1
c ⟶ 1

(B = {1, 2, 3})

is a *constant* function since ∀x ∈ A, f(x) = 1.

Definition 3.1.4

Two functions, f and g, are said to be ***equal*** and we write **f = g** if:

a. $D_f = D_g$

b. $f(a) = g(a), \forall a \in D_f$

It should be noted from the definition of function equality that two conditions must be verified when showing two functions, f and g, are equal:

f and g have the same domain: i.e., $D_f = D_g$; *and*
for each element of the domain, both f and g produce the same image in the range; i.e., $\forall a \in D_f$, $f(a) = g(a)$.

Example 3.1.4

Let A = {a,b,c} and B = {1,2,3}. Consider the following functions f and g where,

f: A ⟶ B	g: A ⟶ B
a ⟶ 1	a ⟶ 1
b ⟶ 3	b ⟶ 2
c ⟶ 2	c ⟶ 2

Now $D_f = D_g = A$. However, for each $x \in A$, f and g do not always produce the same image in the range since f(b) = 3 and g(b) = 2. Hence, $f \neq g$.

Example 3.1.5

Let $f: D_f \to \mathbf{R}$ and $g: D_g \to \mathbf{R}$, where $D_f, D_g \subseteq \mathbf{R}$ and f and g are defined by

a. $f(x) = \frac{x^2 - 4}{x - 2}$ and $g(x) = x + 2$. Then $f \neq g$ since $D_f = \mathbf{R} - \{2\} \neq D_g = \mathbf{R}$; i.e., f and g do not have the same domains.

b. $f(x) = 2x + 1$ and $g(x) = 3x - 1$. Then $D_f = D_g = \mathbf{R}$ but $f \neq g$ since $f(0) \neq g(0)$; i.e., f and g do not always produce the same image in the range for each element in the domain.

c. $f(x) = \frac{1}{x - 2}$ and $g(x) = \frac{4}{x - 2} - \frac{3}{x - 2}$. Then f = g since $D_f = D_g = \mathbf{R} - \{2\}$ and $\forall x \in D_f$, $g(x) = \frac{4}{x - 2} - \frac{3}{x - 2} = \frac{4 - 3}{x - 2} = \frac{1}{x - 2} = f(x)$.

Exercises 3.1

1. Let $A = \{1,2,3\}$ and $B = \{a,b\}$. Exhibit all functions from the set A to the set B (Label the functions f_1, f_2, etc.).

2. Given that $D_f, D_g \subseteq \mathbf{R}$, explain why the following pairs of functions are *not* equal:

   a. $f(x) = \frac{x^2 - 1}{x - 1}$ and $g(x) = x + 1$.

   b. $f(x) = |x|$ and $g(x) = x$.

Section 3.2
TYPES OF FUNCTIONS

Let $f: A \rightarrow B$. It may happen that two or more elements of A have the same image in B. Also, it may happen that there are elements in B that are not the image of any element in A. For example,

f: A ⟶ B		g: A ⟶ B	
a ↘ 2	b ↗ 1	a ⟶ 1	
c ↗ 2		b ↘ 3	2

$f: a \mapsto 2,\ b \mapsto 1,\ c \mapsto 2$; $\quad g: a \mapsto 1,\ b \mapsto 3$ (with 2 not an image).

Functions for which these situations do not occur are of special interest and will be discussed in this section.

Definition 3.2.1

A function f: A $\rightarrow$ B is said to be ***one to one***, and we write **1–1**, if distinct elements in A have distinct images in B. That is, for all $a_1, a_2 \in A$, if $a_1 \neq a_2$, then $f(a_1) \neq f(a_2)$. Functions that are 1–1 are also called ***injective***.

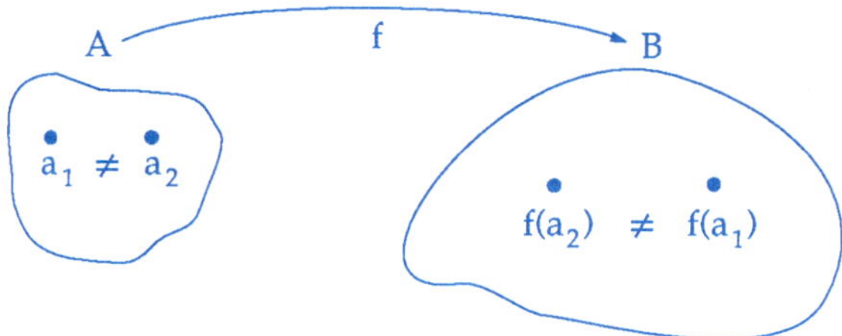

By definition, a function f: A $\rightarrow$ B is 1–1 if $\forall\ a_1, a_2, \in A,\ a_1 \neq a_2 \Rightarrow f(a_1) \neq f(a_2)$. By the contrapositive law, this is logically equivalent to:

$$\sim [f(a_1) \neq f(a_2)] \Rightarrow \sim (a_1 \neq a_2) \text{ or } f(a_1) = f(a_2) \Rightarrow a_1 = a_2.$$

That is, f is 1–1 if, when the elements in the range are the same, then the corresponding domain elements must be the same.

We now have two ways in which to show a function f: A $\rightarrow$ B is 1–1. One arbitrarily selects a_1 , $a_2 \in A$ and shows one of the following propositions is true:

If $a_1 \neq a_2$, then $f(a_1) \neq f(a_2)$; *or*
If $f(a_1) = f(a_2)$, then $a_1 = a_2$.

Example 3.2.1

a. Refer to Example 3.1.1 where A = {a,b,c} and B = {1,2,3}.

f: A $\longrightarrow$ B
a $\rightarrow$ 2
b $\rightarrow$ 1
c $\rightarrow$ 3

g: A $\longrightarrow$ B
a $\rightarrow$ 1
b $\rightarrow$ 1
c $\rightarrow$ 3
(2 has no arrow)

Now f is 1–1 since, whenever $a_1 \neq a_2$, $f(a_1) \neq f(a_2)$. However, g is not 1–1 since $a \neq b$ but $g(a) = g(b) = 1$.

b. Refer to Example 3.1.2, parts (b) and (c).

   i. Here f: $\mathbf{R} \to \mathbf{R}$ where $f(x) = x^2$.
   Now f is not 1–1 since 2, $-2 \in D_f = \mathbf{R}$, and $2 \neq -2$ but $f(2) = f(-2) = 4$.

   ii. Here g: $\mathbf{R} - \{0\} \to \mathbf{R}$ where $g(x) = \frac{1}{x}$. Now g is 1–1.

   Let $a_1, a_2 \in D_g = \mathbf{R} - \{0\}$ and assume $a_1 \neq a_2$.

   We wish to show $g(a_1) \neq g(a_2)$. Now

$$\begin{aligned} a_1 \neq a_2 &\Rightarrow \frac{1}{a_1} \neq \frac{1}{a_2} \\ &\Rightarrow g(a_1) \neq g(a_2) \end{aligned}$$

   Hence, g is 1–1.

c. Let h: $\mathbf{R} \to \mathbf{R}$ where $h(x) = 2x - 1$. Now h is 1–1.

   Let $a_1, a_1 \in D_h = \mathbf{R}$ and assume $h(a_1) = h(a_2)$.

   We wish to show $a_1 = a_2$. Now

$$\begin{aligned} h(a_1) = h(a_2) &\Rightarrow 2a_1 - 1 = 2a_2 - 1 \\ &\Rightarrow 2a_1 = 2a_2 \\ &\Rightarrow a_1 = a_2 \end{aligned}$$

   Hence, h is 1–1.

d. Let h: $\mathbf{N} \times \mathbf{N} \to \mathbf{N}$, where $h((m,n)) = 2^m(2n - 1)$. Now h is 1–1.

   i. Let E be the set of even natural numbers; i.e., $E = \{x \in \mathbf{N} \mid x = 2n, n \in \mathbf{N}\}$ and let O be the set of odd natural numbers; i.e.,
   $O = \{x \in \mathbf{N} \mid x = 2n - 1, n \in \mathbf{N}\}$. If $x \in E$ and $y \in O$, then $xy \in E$.

$$\begin{aligned} xy &= 2t(2s - 1), \text{ where } t, s \in \mathbf{N} \\ &= 2[t(2s - 1)] \\ &= 2n, \text{ where } n = t(2s - 1) \in \mathbf{N} \end{aligned}$$

   Hence, $xy \in E$.

Example 3.2.1 continued next page.

Example 3.2.1 Continued

ii. Let $(m,n), (s,t) \in \mathbf{N} \times \mathbf{N}$ and assume $h((m,n)) = h((s,t))$. We must show $(m,n) = (s,t)$; i.e., $m = s$ and $n = t$. Since $h((m,n)) = h((s,t))$, $2^m(2n - 1) = 2^s(2t - 1)$. Now either $m = s$, $m > s$ or $m < s$. Suppose $m > s$. We then have $2^{m-s}(2n - 1) = 2t - 1$. Since $m - s > 0$, $2^{m-s} \in E$. By part (i) above, $2^{m-s} \in E$ and $2n - 1 \in 0$ yields $2^{m-s}(2n - 1) \in E$. But $2t - 1 \in O$; and, since $2^{m-s}(2n - 1) = 2t - 1$, $2^{m-s}(2n - 1) \in O$. Hence $E \cap O \neq \emptyset$. This is a contradiction since $E \cap O = E \cap E' = \emptyset$. Thus $m \not> s$. In like manner one can show $m \not< s$. Hence, we have $m = s$ which yields

$$\begin{aligned} 2^m(2n - 1) &= 2^m(2t - 1) \\ 2n - 1 &= 2t - 1 \\ 2n &= 2t \\ n &= t \end{aligned}$$

Therefore, $(m,n) = (s,t)$ and h is 1–1.

For every function $f: A \to B$, $R_f \subseteq B$. By Example 3.1.2, parts (a), (b) and (c), we see that we may not assume that the range of a function is always the entire set B. This is due to the fact that not every element in B is necessarily the image of an element in A.

Definition 3.2.2

A function $f: A \to B$ is said to be ***onto*** if each element of B is the image of at least one element of A. That is, for each $b \in B$, there exists at least one $a \in A$ such that $b = f(a)$. Functions that are onto are also called ***surjective***.

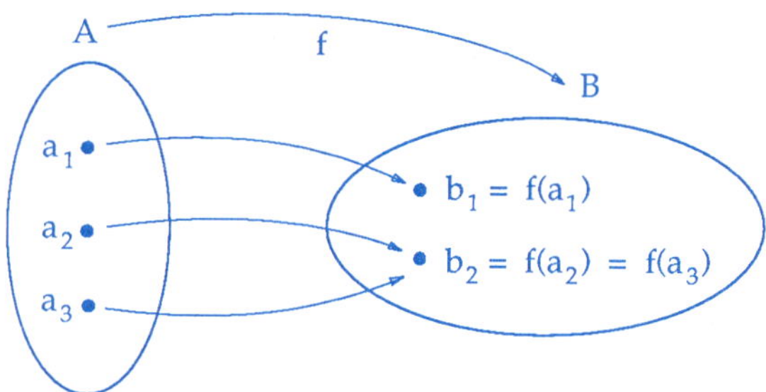

To show that a function f: A ⟶ B is onto one selects an arbitrary element $b \in B$ and shows there exists an element $a \in A$ such that $b = f(a)$. Having done so, one may conclude that f is onto since b represents an arbitrary element of B.

If a function f: A ⟶ B is onto, then for $b \in B$, there is an $a \in A$ such that $b = f(a)$. Hence $b \in R_f$. This shows $B \subseteq R_f$. As stated above $R_f \subseteq B$ is always true for any function f. Therefore, if f is onto, we have $R_f = B$.

Whenever we have a function f: A ⟶ B, we may also consider f as a function from A to R_f; i.e., f: A ⟶ R_f. In this way f becomes a function from A onto R_f.

Example 3.2.2

a. Refer to Example 3.1.1 where A = {a,b,c}, B = {1,2,3} and

f: A ⟶ B
a ⟶ 2
b ⟶ 1
c ⟶ 3

g: A ⟶ B
a ⟶ 1
b ⟶ 1
c ⟶ 3

Now f is onto since for each $y \in B$, there is an $x \in A$ such that $f(x) = y$. However, g is not onto since $2 \in B$ but there is no $x \in A$ such that $g(x) = 2$. Note, however, that g: A ⟶ {1,3} = R_g is onto.

b. Refer to Example 3.1.2, parts (b) and (c).

i. Here f: **R** ⟶ **R** where $f(x) = x^2$. f is not onto since $-3 \in \mathbf{R}$ but there is no $x \in D_f = \mathbf{R}$ such that $f(x) = -3$ or $x^2 = -3$. Note, however, that since $x^2 \geq 0$, $R_f = [0,\infty)$. Therefore, f: **R** ⟶ $[0,\infty) = R_f$ is onto.

ii. Here g: **R** – {0} ⟶ **R** where $g(x) = \frac{1}{x}$. g is not onto since $0 \in \mathbf{R}$ but there is no $x \in D_g = \mathbf{R} - \{0\}$ such that $g(x) = 0$ or $\frac{1}{x} = 0$. Note, however, that if $x \neq 0$, then $\frac{1}{x} \neq 0$. Hence, $\frac{1}{x} \in D_g$ and $g(\frac{1}{x}) = x$. Therefore $R_g = \mathbf{R} - \{0\}$ and g: **R** – {0} ⟶ **R** – {0} = R_g is onto.

Example 3.2.2 continued next page.

Example 3.2.2 Continued

c. Let h: $\mathbf{R} \to \mathbf{R}$ where $h(x) = 2x - 1$. We wish to show h is onto. Hence, we select an arbitrary real number $y \in \mathbf{R}$ and show there exists at least one real number x such that $h(x) = y$. Now if $h(x) = y$, then

$$\begin{aligned} h(x) = y &\Rightarrow 2x - 1 = y \\ &\Rightarrow 2x = y + 1 \\ &\Rightarrow x = \frac{y+1}{2} \end{aligned}$$

Since $y \in \mathbf{R}$, $x = \frac{y+1}{2} \in \mathbf{R}$ and we have

$$\begin{aligned} h(x) &= h(\tfrac{y+1}{2}) \\ &= 2(\tfrac{y+1}{2}) - 1 \\ &= (y + 1) - 1 \\ &= y \end{aligned}$$

Therefore, h is onto.

d. Refer to Example 3.2.1, part (d). Here h: $\mathbf{N} \times \mathbf{N} \to \mathbf{N}$ where $h((m,n)) = 2^m(2n - 1)$. Now h is not onto since, $\forall (m,n) \in N \times N$, $h((m,n)) \in E$; i.e., $h((m,n))$ is an even natural number. Thus, if $x \in \mathbf{N}$ where x is odd, there does not exist $(m,n) \in \mathbf{N} \times \mathbf{N}$ such that $h((m,n)) = x$. Note, however, that h: $\mathbf{N} \times \mathbf{N} \to E = R_h$ is onto.

e. Let g: $A \times B \to B \times A$ where $g((a,b)) = (b,a)$ and $A \neq \emptyset$ and $B \neq \emptyset$. We wish to show that g is onto. Hence, we select an arbitrary element $(b,a) \in B \times A$ and show there exists at least one element $(x,y) \in A \times B$ such that $g((x,y)) = (b,a)$. Now if $g((x,y)) = (b,a)$, then

$$\begin{aligned} g((x,y)) = (b,a) &\Rightarrow (y,x) = (b,a) \\ &\Rightarrow y = b \text{ and } x = a \end{aligned}$$

Since $a \in A$ and $b \in B$, $(x,y) = (a,b) \in A \times B$ and we have

$$\begin{aligned} g((x,y)) &= g((a,b)) \\ &= (b,a) \end{aligned}$$

Therefore, g is onto.

In many instances we wish to consider special functions f: A ⟶ B in which each element of B is both the image of at least one element of A and the image of at most one element of A. This leads us to the following definition.

Definition 3.2.3

A function f: A ⟶ B is called a ***one to one correspondence***, and we write f is a **1–1** ***correspondence***, if f is both 1–1 and onto. Functions that are 1–1 and onto are also called ***bijective***.

If the function f: A ⟶ B is a 1–1 *correspondence*, then one may intuitively think of the sets A and B as being of the same "size" or as having the same "number of elements."

Example 3.2.3

a. Refer to Example 3.1.1 where A = {a,b,c}, B = {1,2,3} and

f: A ⟶ B

a ↘↗ 1
b ↗↘ 2
c ⟶ 3

f is a 1–1 correspondence since f is 1–1 and onto.

b. Refer to Examples 3.2.1 and 3.2.2, part (c), where h: **R** ⟶ **R** and h(x) = 2x – 1. We have already shown h is 1–1 and onto. Hence h is a 1–1 correspondence.

c. In Example 3.2.2, part (e), we have already shown the function g: A x B ⟶ B x A where g(a,b) = (b,a) is onto. Let us now show g is 1–1. Let $(a_1,b_1), (a_2,b_2) \in A \times B$ and assume $g((a_1,b_1)) = g((a_2,b_2))$. We must show $(a_1,b_1) = (a_2,b_2)$. Now

$$\begin{aligned} g((a_1,b_1)) = g((a_2,b_2)) &\Rightarrow (b_1,a_1) = (b_2,a_2) \\ &\Rightarrow b_1 = b_2 \text{ and } a_1 = a_2 \end{aligned}$$

Thus $(a_1,b_1) = (a_2,b_2)$ and g is 1–1. Since g is 1–1 and onto, g is a 1–1 correspondence.

Let f: A ⟶ B. Then f: A ⟶ $R_f \subseteq$ B is always an onto function. Thus, if f: A ⟶ B is 1–1, then f: A ⟶ R_f is a 1–1 correspondence.

Example 3.2.4

a. Let A = {a,b,c} and B = {1,2,3,4} where

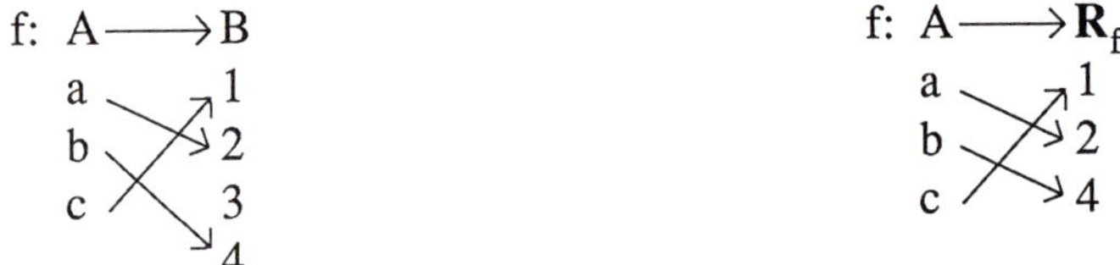

Now f: A ⟶ B is 1–1 but not onto. However, if we consider f: A ⟶ R_f = {f(a), f(b), f(c)} = {2,4,1}, then we see that f is 1–1 and onto and hence a 1–1 correspondence.

b. Refer to Examples 3.2.1 and 3.2.2, part (b), ii, where g: **R** – {0} ⟶ **R** and $g(x) = \frac{1}{x}$. Now g is not a 1–1 correspondence since g is 1–1 but not onto. However, if we consider g: **R** – {0} ⟶ R_g = **R** – {0}, then g is a 1–1 correspondence.

c. Recall from Examples 3.2.1 and 3.2.2, part (d), the function h: **N** x **N** ⟶ **N** where $h((m,n)) = 2^m(2n - 1)$. We showed that h is 1–1 but not onto. However, if we consider h as a function from **N** x **N** to E, the set of even natural numbers, then h is onto. Thus, the function h: **N** x **N** ⟶ E is a 1–1 correspondence.

Theorem 3.2.1

If A has m elements and B has n elements, then A x B has mn elements where m, n ∈ **N**.

Proof

Let A have m elements and B have n elements. Now B can be represented as $B = \{b_1, b_2, \ldots, b_n\}$.

For each i, $i = 1, 2, \ldots n$, form the product set $A \times \{b_i\} = \{(a,b_i) \mid a \in A\}$. The functions $f_i : A \to A \times \{b_i\}$ where $f_i(a) = (a,b_i)$, $i = 1, 2, \ldots, n$, are each 1–1 correspondences. (See Exercises 3.2, problem (5).) Hence, for each i, $i = 1, 2, \ldots, n$, A and $A \times \{b_i\}$ have the same number of elements; i.e., m.

By Theorem 2.5.3, part (a),

$$\bigcup_{i=1}^{n} (A \times \{b_i\}) = A \times \left(\bigcup_{i=1}^{n} \{b_i\}\right) = A \times B.$$

For each i, $i = 1, 2, \ldots, n$, $A \times \{b_i\}$ has m elements; and, by Theorem 2.5.3, part (b), for $i \neq j$,

$$(A \times \{b_i\}) \cap (A \times \{b_j\}) = A \times (\{b_i\} \cap \{b_j\}) = A \times \emptyset = \emptyset.$$

Hence, since $A \times B = (A \times \{b_1\}) \cup (A \times \{b_2\}) \cup \ldots \cup (A \times \{b_n\})$, $A \times B$ has $m + m + \ldots + m$ (n times) or mn elements. ■

Exercises 3.2

1. In problem (1) from Exercises 3.1, you exhibited all functions from the set $A = \{1,2,3\}$ to the set $B = \{a,b\}$.
   a. Identify those functions which are onto.
   b. Can any of the functions be 1–1? Explain.
   c. If A has m elements and B has n elements, how many different functions are there from A to B?

2. Let $f: \mathbf{R} \to \mathbf{R}$ where $f(x) = 3x + 5$. Show that the function f is 1–1 and onto.

3. Let $f: \mathbf{R} \to \mathbf{R}$ where

$$f(x) = \begin{cases} x + 2 & \text{when } x \leq -1 \\ -x & \text{when } -1 < x < 1 \\ x - 2 & \text{when } x \geq 1 \end{cases}$$

   Show that the function f is onto but not 1–1. (A graph of f may be helpful.)

Exercises 3.2 continued next page.

4. a. Let $A = \{1,2,3\}$ and $B = \{x,y\}$. Find $A \times B$ and by means of arrows exhibit the function $f: A \times B \to A$ where $f((a, b)) = a$.

   b. Let A and B be any two nonempty sets and $f: A \times B \to A$ where $f((a,b)) = a$. (The function f is called the **projection** of $A \times B$ onto A.)
      i. Show that the function f is onto.
      ii. Is the function f necessarily 1–1? (Support your answer either by means of a proof or a counter-example.)

5. Let $A \neq \emptyset$ and b a fixed element. Show that the function $f: A \to A \times \{b\}$ where $f(a) = (a,b)$ is a 1–1 correspondence.

6. Let A, B and C be non-empty sets. Show that the function $f: (A \times B) \times C \to A \times B \times C$ where $f((a,b), c) = (a,b,c)$ is a 1–1 correspondence and hence $(A \times B) \times C$ and $A \times B \times C$ have the same number of elements. (See Exercises 2.5, problem (5), for the definition of $A \times B \times C$.)

7. Let B be a set and $n \in \mathbf{N}$. We define $B^n = B \times B \times \ldots \times B = \{(b_1,b_2, \ldots ,b_n) \mid b_i \in B\}$. (See Exercises 2.5, problem (5), where $A_i = B$, $i = 1, 2, \ldots ,n$). If $B = \{0,1\}$, show B^n has 2^n elements. (Hint: Consider a proof by induction. For $n \geq 1$, let p(n): If $B = \{0,1\}$, then B^n has 2^n elements. Use the results of Theorem 3.2.1 and a generalization of problem (6) above in showing $p(k + 1)$ is true.)

8. a. Let $A = \{a_1,a_2,a_3\}$. List the elements of P(A).
   b. Let $B = \{0,1\}$. List the elements of $B^3 = B \times B \times B$.
   c. Consider the function $f: P(A) \to B^3$ where for $X \in P(A)$, $f(X) = (b_1,b_2,b_3)$ and, for $i = 1, 2, 3$,

$$b_i = \begin{cases} 0 & \text{if } a_i \notin X \\ 1 & \text{if } a_i \in X \end{cases}$$

   By means of arrows exhibit the function $f: P(A) \to B^3$.

9. Show that if A has n elements, then the power set of A, P(A), has 2^n elements. (Hint: Let $A = \{a_1, a_2, \ldots , a_n\}$, $B = \{0,1\}$ and consider the function $f: P(A) \to B^n$ where for $X \in P(A)$, $f(X) = (b_1, b_2, \ldots , b_n)$; and, for $i = 1, 2, \ldots ,n$,

$$b_i = \begin{cases} 0 & \text{if } a_i \notin X \\ 1 & \text{if } a_i \in X \end{cases}$$

   Show f is a 1–1 correspondence and hence P(A) and B^n have the same number of elements; i.e., 2^n.

Section 3.3
CONSTRUCTION OF FUNCTIONS

It is sometimes desirable to construct new functions from given functions. This can be accomplished using various techniques provided certain conditions are met. In this section we shall limit our discussion to three of the more important techniques of constructing new functions.

One technique is to construct a new function from a given function by adding to (extending) or deleting from (restricting) the given function's domain.

Definition 3.3.1

Let $X \subseteq A$, where $X \neq \emptyset$, $f: A \rightarrow B$ and $g: X \rightarrow B$.

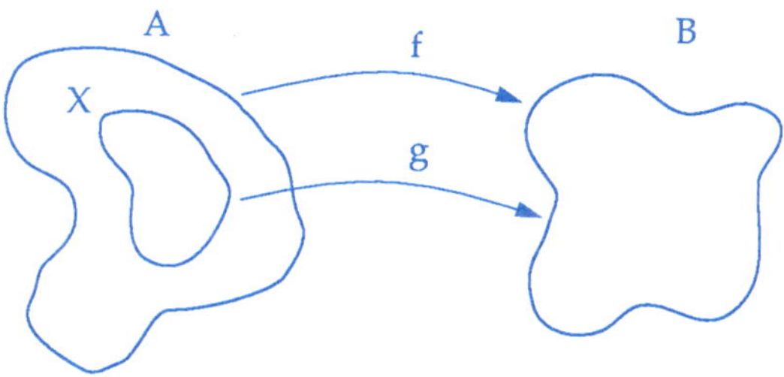

If $\forall x \in X$, $g(x) = f(x)$, then g is called a ***restriction*** of f to X and f is called an ***extension*** of g to A. In either case we write $\mathbf{g = f\,|_X}$.

Let $f: A \rightarrow B$. If g is a restriction of f to a non-empty subset X of A, then the image of any element in X under g is the same as its image under f. In like manner, if h is an extension of f to a superset Y of A, then the image of each element of A under h must be the same as its image under f.

Example 3.3.1

a. Let A = {a,b,c}, X = {a,c} and B = {1,2,3,4}. Consider the functions f: A ⟶ B and g: X ⟶ B where

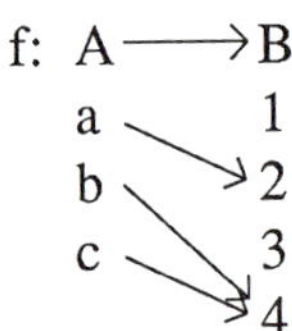

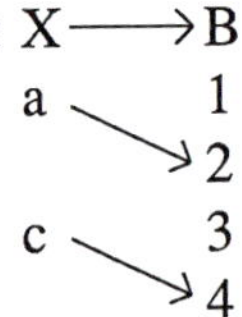

Since $X \subseteq A$, $X \neq \emptyset$ and $g(x) = f(x)$, $\forall x \in X$, g is a restriction of f to X and f is an extension of g to A; i.e., $g = f|_X$.

b. Let f: **R** ⟶ **R** where $f(x) = x + 2$ and g: **R** – {2} ⟶ **R** where $g(x) = \frac{x^2 - 4}{x - 2}$.

Now **R** – {2} ⊂ **R** and $\forall x \in$ **R** – {2},

$$\begin{aligned} g(x) &= \frac{x^2 - 4}{x - 2} \\ &= \frac{(x-2)(x+2)}{x - 2} \\ &= x + 2 \\ &= f(x) \end{aligned}$$

Thus, g is a restriction of f to **R** – {2} and f is an extension of g to **R**; i.e., $g = f|_{\mathbf{R} - \{2\}}$.

c. Let A = {b,d}, B = {1,2,3,4} and f: A ⟶ B where

f: A ⟶ B
b ⟶ 1
2
d ⟶ 3
4

If Y = {a,b,c,d}, then A ⊂ Y. We wish to construct a function h: Y ⟶ B such that h is an extension of f to Y. Keep in mind that h must satisfy the condition that $\forall x \in A$, $h(x) = f(x)$; i.e., $h(b) = f(b) = 1$ and $h(d) = f(d) = 3$. We thus have the partial correspondence on the next page:

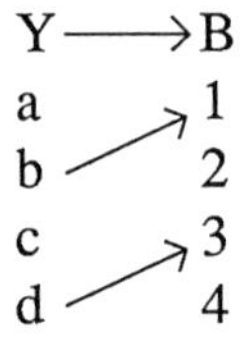

$$
\begin{array}{ccc}
Y & \longrightarrow & B \\
a & & 1 \\
b & \nearrow & 2 \\
c & & 3 \\
d & \nearrow & 4
\end{array}
$$

While there are several ways in which h can be defined as a function from Y to B, consider the function h: Y → B where

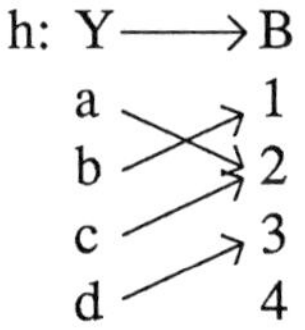

$$
\begin{array}{lcl}
h: Y & \longrightarrow & B \\
a & \searrow & 1 \\
b & \nearrow & 2 \\
c & \nearrow & 3 \\
d & \nearrow & 4
\end{array}
$$

Thus h is an extension of f to the superset Y, i.e., $f = h|_A$.

d. Let g: **Z** → **Z** where g(n) = n. Since **Z** ⊂ **R**, we wish to construct a function f: **R** → **Z** such that f is an extension of g to **R**. Keep in mind f must satisfy the condition that ∀n ∈ **Z**, f(n) = g(n). If x ∈ **R**, then there exists n ∈ **Z** such that $n \le x < n + 1$. Consider the function f: **R** → **Z** where, if $n \le x < n + 1$, then f(x) = n. Then if n ∈ **Z**, f(n) = n = g(n). Thus f is an extension of g to **R**; i.e., $g = f|_{\mathbf{Z}}$.

e. Let g: **N** → **N** where g(x) = 2x – 1. Since **N** ⊂ **Z**, we wish to construct a function f: **Z** → **N** such that f is an extension of g to **Z**. Keep in mind f must satisfy the condition that ∀x ∈ **N**, f(x) = g(x). Consider the piece-wise function f: **Z** → **N** where

$$f(x) = \begin{cases} |x| + 1 \text{ when } x \in \mathbf{Z} - \mathbf{N} \\ 2x - 1 \text{ when } x \in \mathbf{N} \end{cases}$$

Then if x ∈ **N**, f(x) = 2x – 1 = g(x). Thus f is an extension of g to **Z**; i.e., $g = f|_{\mathbf{N}}$.

One can also construct a new function from given functions by combining the domains of the given functions using the set operation of union.

Definition 3.3.2

Let f: $D_f \rightarrow B$ and g: $D_g \rightarrow C$. If $\forall x \in D_f \cap D_g$, f(x) = g(x), then the ***union*** of f and g, denoted **f ∪ g**, is the function f ∪ g: $D_{f \cup g} \rightarrow B \cup C$ where $D_{f \cup g} = D_f \cup D_g$ and

$$(f \cup g)(x) = \begin{cases} f(x) & \text{when } x \in D_f - D_g \\ f(x) = g(x) & \text{when } x \in D_f \cap D_g \\ g(x) & \text{when } x \in D_g - D_f. \end{cases}$$

Note that in order for f ∪ g to be defined it is essential that f(x) = g(x), $\forall x \in D_f \cap D_g$. Also note that f ∪ g is an extension of both f and g to $D_{f \cup g}$.

Example 3.3.2

a. Let f: $[0,\infty) \rightarrow \mathbf{R}$ where f(x) = 2x and g: $(-\infty,0] \rightarrow \mathbf{R}$ where g(x) = 3x. Now $D_f \cap D_g = \{0\}$ and f(0) = g(0). Hence, f ∪ g is defined. Now $D_{f \cup g} = D_f \cup D_g = \mathbf{R}$ andf ∪ g: $\mathbf{R} \rightarrow \mathbf{R}$ where

$$(f \cup g)(x) = \begin{cases} 2x & \text{when } x > 0 \ (x \in D_f - D_g = (0,\infty)) \\ 0 & \text{when } x = 0 \ (x \in D_f \cap D_g = \{0\}) \\ 3x & \text{when } x < 0 \ (x \in D_g - D_g = (-\infty,0)) \end{cases}$$

b. Let f: $(-\infty,0] \rightarrow \mathbf{R}$ where f(x) = x and g: $(0,\infty) \rightarrow \mathbf{R}$ where $g(x) = \frac{1}{x}$. Since $D_f \cap D_g = \emptyset$, f ∪ g is defined. Now $D_{f \cup g} = D_f \cup D_g = \mathbf{R}$ and f ∪ g: $\mathbf{R} \rightarrow \mathbf{R}$ where

$$(f \cup g)(x) = \begin{cases} x & \text{when } x \leq 0 \ (x \in D_f - D_g = D_f) \\ \frac{1}{x} & \text{when } x > 0 \ (x \in D_g - D_f = D_g) \end{cases}$$

c. Let f: $[0,\infty) \rightarrow \mathbf{R}$ where f(x) = 2x + 1 and g: $(-\infty,0] \rightarrow \mathbf{R}$ where g(x) = 3x. Here f ∪ g is not defined since $D_f \cap D_g = \{0\}$ but f(0) ≠ g(0).

d. Let f: $\mathbf{R} \rightarrow \mathbf{R}$ be defined by the constant function f(x) = 0 and g: $\mathbf{R} \rightarrow \mathbf{R}$ where g(x) = sin x. Now f ∪ g is not defined since $D_f \cap D_g = \mathbf{R}$ but, except for the set $\{k\pi \mid k \in \mathbf{Z}\}$, f(x) ≠ g(x).

The third technique used in the construction of a new function from given functions is referred to as the *composition of* functions. Let f: A ⟶ B and g: B ⟶ C be two functions. If a ∈ A, then f(a) ∈ B. Since f(a) ∈ B = D_g, g(f(a)) is defined and g(f(a)) ∈ C. Thus, a correspondence exists between the elements of A and the elements of C. Since f and g are functions, f(a) is an unique element of B and g(f(a)) is an unique element of C. Hence, for each a ∈ A there is an unique element g(f(a)) ∈ C and the correspondence from A to C is a function.

Definition 3.3.3

Let f: A ⟶ B and g: B ⟶ C be two functions. The ***composition*** of g and f, denoted by **g ∘ f** (read "g circle f"), is the function g ∘ f: A ⟶ C defined by $(g \circ f)(a) = g(f(a)), \forall a \in A$.

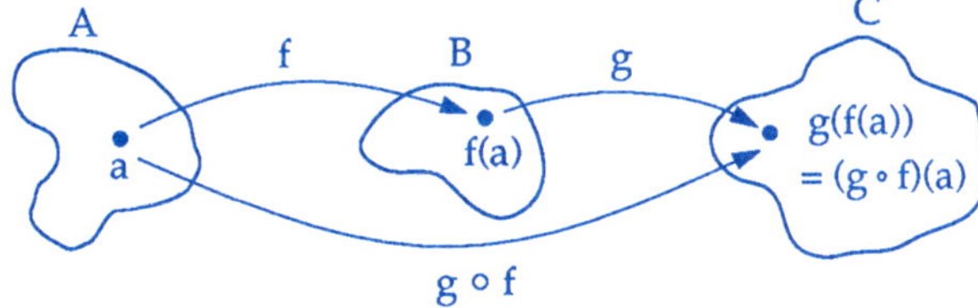

Note that $D_{g \circ f} = D_f = A$.

Example 3.3.3

a. Let f, g be defined by:

f: A ⟶ B
a ⟶ 2
b ⟶ 3
c ⟶ 1

g: B ⟶ C
1 ⟶ y
2 ⟶ x
3 ⟶ y
z

Then $(g \circ f)(a) = g(f(a)) = g(2) = x$

$(g \circ f)(b) = g(f(b)) = g(3) = y$

$(g \circ f)(c) = g(f(c)) = g(1) = y$

Thus g ∘ f: A ⟶ C
a ⟶ x
b ⟶ y
c ⟶ y
z

Example 3.3.3 continued next page.

b. Let f: **R** → **R** where f(x) = 2x – 1 and g: **R** → **R** where $g(x) = x^2$.

i. Then g ∘ f: **R** → **R** where:

$$\begin{aligned}(g \circ f)(x) &= g(f(x)) \\ &= g(2x - 1) \\ &= (2x - 1)^2 \\ &= 4x^2 - 4x + 1 \text{ and } D_{g \circ f} = D_f = \mathbf{R}.\end{aligned}$$

ii. Then f ∘ g: **R** → **R** where:

$$\begin{aligned}(f \circ g)(x) &= f(g(x)) \\ &= f(x^2) \\ &= 2(x^2) - 1 \\ &= 2x^2 - 1 \text{ and } D_{f \circ g} = D_g = \mathbf{R}.\end{aligned}$$

iii. Parts (i) and (ii) show that in general $f \circ g \neq g \circ f$.

Theorem 3.3.1

Let f: A → B, g: B → C and h: C → D.
Then $(h \circ g) \circ f = h \circ (g \circ f)$

Proof

We wish to show the functions $(h \circ g) \circ f$ and $h \circ (g \circ f)$ are equal. From the definition of function equality, we must show the following two conditions hold:

$D_{(h \circ g) \circ f} = D_{h \circ (g \circ f)}$; and
$((h \circ g) \circ f)(a) = (h \circ (g \circ f))(a)$, $\forall a \in D_{(h \circ g) \circ f}$.

$D_{(h \circ g) \circ f} = D_f = A$ and $D_{h \circ (g \circ f)} = D_{g \circ f} = D_f = A$.
Thus, $D_{(h \circ g) \circ f} = D_{h \circ (g \circ f)} = A$.

Let $a \in A$. Then

$$\begin{aligned}((h \circ g) \circ f)(a) &= (h \circ g)(f(a)) \\ &= h(g(f(a))) \\ &= h((g \circ f)(a)) \\ &= (h \circ (g \circ f))(a)\end{aligned}$$

Therefore, $\forall a \in A$, $((h \circ g) \circ f)(a) = (h \circ (g \circ f))(a)$.

Hence, $(h \circ g) \circ f = h \circ (g \circ f)$. ■

Theorem 3.3.2

Let f: A → B and g: B → C.

a. If f and g are 1–1, then $g \circ f$ is 1–1.
b. If f and g are onto, then $g \circ f$ is onto.

Proof

a. Let f and g be 1–1 and $a_1, a_2 \in A$ where $a_1 \neq a_2$. We must show $(g \circ f)(a_1) \neq (g \circ f)(a_2)$. Since f: A → B is 1–1. $f(a_1) \neq f(a_2)$. Now $f(a_1), f(a_2) \in B$ and g: B → C is 1–1, therefore $g(f(a_1)) \neq g(f(a_2))$ or $(g \circ f)(a_1) \neq (g \circ f)(a_2)$. Hence, $g \circ f$ is 1–1.

■

b. Let f and g be onto and select an arbitrary element $c \in C$. To show $g \circ f$ is onto we must show $\exists a \in A$ such that $(g \circ f)(a) = c$. Since g is onto $\exists b \in B$ such that $g(b) = c$. Also, since f is onto and $b \in B$, $\exists a \in A$ such that $f(a) = b$. Therefore, $a \in A$ and $(g \circ f)(a) = g(f(a)) = g(b) = c$. Hence, $g \circ f$ is onto.

■

Corollary

Let f: A → B and g: B → C. If f and g are both 1–1 correspondences, then $g \circ f$ is a 1–1 correspondence.

Exercises 3.3

1. Let g: **N** → **R** where $g(x) = x^2$. Find three different functions f_i: **R** → **R**, i = 1, 2, 3, such that each f_i is an extension of g to **R**; i.e., $g = f_i|_{\mathbf{N}}$.

2. Let f: $(-\infty,1]$ → **R** where $f(x) = x^2$ and g: $[1,\infty)$ → **R** where $g(x) = 2 - x$. Find the function $f \cup g$ if it exists.

3. Let f: $(-\infty,0]$ → **R** where $f(x) = -x$ and g: **R** → **R** where $g(x) = |x|$. Find the function $f \cup g$ if it exists.

4. For each pair of functions, f and g, find $(f \circ g)(x)$ and $(g \circ f)(x)$. Assume $D_f, D_g \subseteq \mathbf{R}$.
   a. $f(x) = x^2 - 3x$ and $g(x) = 2x + 5$.
   b. $f(x) = \frac{x + 2}{x - 3}$ and $g(x) = x^2 - 1$.

5. Let f: A → B. Prove $f \circ \mathbf{e}_A = f$ and $\mathbf{e}_B \circ f = f$ where $\mathbf{e}_A$ is the identity function on A and $\mathbf{e}_B$ is the identity function on B.

6. Let f: A → B, g: B → C and $g \circ f$: A → C.
   a. If the function $g \circ f$ is onto, prove that the function g is onto.
   b. If the function $g \circ f$ is 1–1, prove that the function f is 1–1.

Section 3.4
INVERSE FUNCTIONS

Let f: A ⟶ B be a function. We would now like to consider reversing the correspondence given by the function f. That is, if a ⟶ b under f, we would like to consider the correspondence in the reverse direction; i.e., b ⟶ a. The question arises as to whether or not the reversing of the correspondence given by f yields a function. Consider the following examples of functions and their respective reverse correspondences:

f: A	⟶	B	g: A	⟶	B	h: A	⟶	B
a	⟶	1	a		1	a		1
b		2	b		2	b		2
c		3	c		3	c		3
					4	d		

B	⟶	A	B	⟶	A	B	⟶	A
1	⟶	a	1		a	1		a
2		b	2		b	2		b
3		c	3		c	3		c
			4					d

We see that for function f the reverse correspondence from B to A is a function, but that the reverse correspondences for g and h do not yield functions. Since g is **not onto**, the reverse correspondence for g cannot make an assignment for $1 \in B$. Since h is **not 1–1**, the reverse correspondence for h assigns the element $2 \in B$ to the elements $a \in A$ and $c \in A$.

Definition 3.4.1

Let f: A ⟶ B. If reversing the correspondence given by f yields a function from B to A, we shall call the correspondence the ***inverse function of f*** and denote the inverse function by the symbol $\mathbf{f^{-1}}$.

Suppose f: A $\longrightarrow$ B and f^{-1}: B $\longrightarrow$ A.

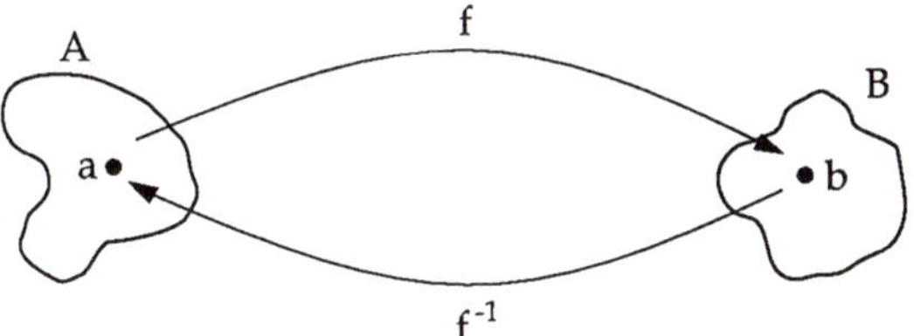

$$f: a \longrightarrow b = f(a) \text{ and } f^{-1}: b \longrightarrow a = f^{-1}(b).$$

Hence, $a = f^{-1}(b) \Leftrightarrow b = f(a)$.

Theorem 3.4.1

Let f: A $\longrightarrow$ B. If f^{-1} exists, then $f^{-1} \circ f = \mathbf{e}_A$ and $f \circ f^{-1} = \mathbf{e}_B$ where $\mathbf{e}_A$ is the identity function on set A and $\mathbf{e}_B$ is the identity function on set B.

Proof

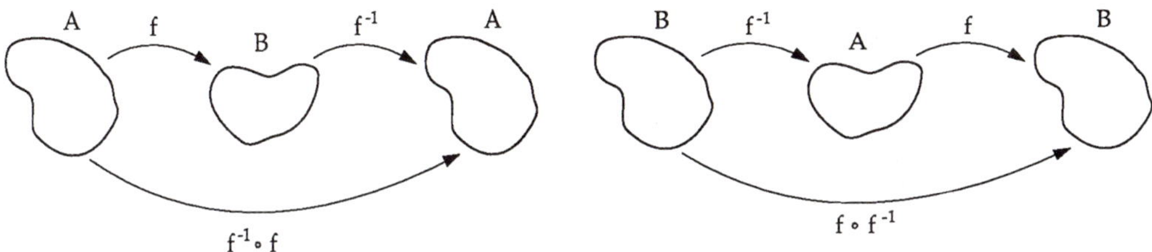

We shall prove $f^{-1} \circ f = \mathbf{e}_A$ and leave the proof of $f \circ f^{-1} = \mathbf{e}_B$ as an exercise. Let f: A $\longrightarrow$ B and f^{-1} exist, where f^{-1}: B $\longrightarrow$ A. We must show the functions $f^{-1} \circ f$ and $\mathbf{e}_A$, where $\mathbf{e}_A(a) = a$, $\forall a \in A$, are equal.

$D_{f^{-1} \circ f} = D_f = A$ and $D_{\mathbf{e}_A} = A$. Hence $D_{f^{-1} \circ f} = D_{\mathbf{e}_A}$.

Let $a \in A$ and $b = f(a)$. Since f^{-1} exists, $f^{-1}(b) = a$. Therefore,

$$\begin{aligned}(f^{-1} \circ f)(a) &= f^{-1}(f(a)) \\ &= f^{-1}(b) \\ &= a \\ &= \mathbf{e}_A(a)\end{aligned}$$

Hence $f^{-1} \circ f = \mathbf{e}_A$. ■

The function f: A ⟶ B given at the beginning of this section whose reverse correspondence yields a function indicates a condition that guarantees the existence of an inverse function; i.e., that of being a 1–1 correspondence. This result is stated formally in the following theorem.

Theorem 3.4.2

Let f: A ⟶ B. If f is a 1–1 correspondence, then

a. f^{-1} exists
b. f^{-1}: B ⟶ A is a 1–1 correspondence.

Proof

a. Let f: A ⟶ B be a 1–1 correspondence and b ∈ B. Since f is onto, there exists a ∈ A such that b = f(a). Since f is 1–1, a is the only element of A such that b = f(a). Hence, each element b ∈ B can be written in the form f(a), where a ∈ A, in one and only one way. Thus, if a ⟶ b under f, the reverse correspondence, b ⟶ a, is a function from B ⟶ A. Thus f^{-1} exists. ■

b. Let f: A ⟶ B be a 1–1 correspondence and f^{-1}: B ⟶ A. We must show f^{-1} is 1–1 and onto.

i. f^{-1} is 1–1:

Let b_1, b_2, ∈ B. Assume $f^{-1}(b_1) = f^{-1}(b_2)$. Since f: A ⟶ B and $f^{-1}(b_1), f^{-1}(b_2) \in A$, $f(f^{-1}(b_1)) = f(f^{-1}(b_2))$ or $(f \circ f^{-1})(b_1) = (f \circ f^{-1})(b_2)$. By Theorem 3.4.1, $f \circ f^{-1} = \mathbf{e}_B$ where $\mathbf{e}_B(b) = b$, ∀b ∈ B. Hence, we have $\mathbf{e}_B(b_1) = \mathbf{e}_B(b_2)$ which yields $b_1 = b_2$. Thus, f^{-1} is 1–1.

ii. f^{-1} is onto:

Let a ∈ A. We must show there exists b ∈ B such that $f^{-1}(b) = a$. Since a ∈ A and f: A ⟶ B, f(a) ∈ B. Let b = f(a). Then

$$\begin{aligned} f^{-1}(b) &= f^{-1}(f(a)) \\ &= (f^{-1} \circ f)(a) \\ &= \mathbf{e}_A(a) \\ &= a \end{aligned}$$

Hence, f^{-1} is onto.

Therefore, f^{-1} is a 1–1 correspondence. ■

Let $f: A \to B$ such that f is a 1–1 correspondence. Then $D_f = A$ and $R_f = B$. Since $f^{-1}: B \to A$ exists and is also a 1–1 correspondence, $D_{f^{-1}} = B$ and $R_{f^{-1}} = A$. Hence, we see $D_{f^{-1}} = R_f$ and $R_{f^{-1}} = D_f$.

f: A	⟶	B		f^{-1}: B	⟶	A
a	⤨	1		1	⤨	a
b		2		2		b
c	⟶	3		3	⟶	c

$D_{f^{-1}} = R_f = B$ and $R_{f^{-1}} = D_f = A$.

Example 3.4.1

Recall that if $f: A \to B$ is 1–1, then $f: A \to R_f$ is a 1–1 correspondence. Hence, $f^{-1}: R_f \to A$ exists.

a. Let $f: [0,\infty) \to \mathbf{R}$ where $f(x) = \sqrt{x}$. Now f is 1–1, $R_f = [0,\infty)$ and hence $f: [0,\infty) \to [0,\infty)$ is a 1–1 correspondence. Thus $f^{-1}: [0,\infty) \to [0,\infty)$ exists. To find $f^{-1}(x)$, select $x \in D_{f^{-1}} = [0,\infty)$. Then $f^{-1}(x) \in R_{f^{-1}} = [0,\infty)$. Let $f^{-1}(x) = y$. Since $f^{-1}(x) = y$, $x = f(y) = \sqrt{y}$. Solving the equation for y, we get $y = x^2$. But $y = f^{-1}(x)$. Hence, $f^{-1}(x) = x^2$.

b. Let $g: [-\frac{\pi}{2}, \frac{\pi}{2}] \to \mathbf{R}$ where $g(x) = \sin x$. Now g is 1–1, $R_g = [-1,1]$ and hence $g: [-\frac{\pi}{2}, \frac{\pi}{2}] \to [-1,1]$ is a 1–1 correspondence. Thus $g^{-1}: [-1,1] \to [-\frac{\pi}{2}, \frac{\pi}{2}]$ exists. Select $x \in D_{g^{-1}} = [-1,1]$. Then $g^{-1}(x) \in R_{g^{-1}} = [-\frac{\pi}{2}, \frac{\pi}{2}]$. Let $g^{-1}(x) = y$. Since $g^{-1}(x) = y$, $x = g(y) = \sin y$. Solving the equation for y, we get $y = \sin^{-1} x$. But $y = g^{-1}(x)$. Hence $g^{-1}(x) = \sin^{-1} x$.

c. Let $h: \mathbf{R} \to \mathbf{R}$ where $h(x) = 2x - 1$. Now we showed h is a 1–1 correspondence. Hence, $h^{-1}: \mathbf{R} \to \mathbf{R}$ exists. Select $x \in D_{h^{-1}} = \mathbf{R}$. Then $h^{-1}(x) \in R_{h^{-1}} = \mathbf{R}$. Let $h^{-1}(x) = y$. Since $h^{-1}(x) = y$, $x = h(y) = 2y - 1$. Solving the equation for y, we get $y = \frac{x+1}{2}$. But $y = h^{-1}(x)$. Hence, $h^{-1}(x) = \frac{x+1}{2}$.

d. Let $g: A \times B \to B \times A$ where $A \neq \emptyset$, $B \neq \emptyset$ and $g((a,b)) = (b,a)$. We showed g is a 1–1 correspondence. Hence $g^{-1}: B \times A \to A \times B$ exists. Select $(b,a) \in D_{g^{-1}} = B \times A$. Then $g^{-1}((b,a)) \in R_{g^{-1}} = A \times B$. Let $g^{-1}((b,a)) = (x,y)$. Since $g^{-1}((b,a)) = (x,y)$, $(b,a) = g((x,y)) = (y,x)$. Thus, $y = b$ and $x = a$ and therefore $(x,y) = (a,b)$. But $(x,y) = g^{-1}((b,a))$. Hence, $g^{-1}((b,a)) = (a,b)$.

Theorem 3.4.3

Let f: $A \to B$ and g: $B \to C$ where both f and g are 1–1 correspondences. Then $(g \circ f)^{-1} = f^{-1} \circ g^{-1}$.

Proof

Let f: $A \to B$ and g: $B \to C$ be both 1–1 correspondences. By Theorem 3.4.2, f^{-1} and g^{-1} exist. By Corollary to Theorem 3.3.2, $g \circ f$ is a 1–1 correspondence and hence by Theorem 3.4.2, $(g \circ f)^{-1}$ exists. We now wish to show the functions $(g \circ f)^{-1}$ and $f^{-1} \circ g^{-1}$ are equal.

$D_{(g \circ f)^{-1}} = R_{g \circ f} = C$ and $D_{f^{-1} \circ g^{-1}} = D_{g^{-1}} = R_g = C$. Hence, $D_{(g \circ f)^{-1}} = R_{f^{-1} \circ g^{-1}}$.

Let $c \in C = D_{(g \circ f)^{-1}}$. We wish to show $(g \circ f)^{-1}(c) = (f^{-1} \circ g^{-1})(c)$. Since g^{-1} exists, there exists an unique $b \in B$ such that $g^{-1}(c) = b$. Also, since f^{-1} exists, there exists an unique $a \in A$ such that $f^{-1}(b) = a$. Hence, $(f^{-1} \circ g^{-1})(c) = f^{-1}(g^{-1}(c)) = f^{-1}(b) = a$. Now, $g^{-1}(c) = b \Leftrightarrow g(b) = c$ and $f^{-1}(b) = a \Leftrightarrow f(a) = b$. Thus, $(g \circ f)(a) = g(f(a)) = g(b) = c$. Since $(g \circ f)^{-1}$ exists, $(g \circ f)^{-1}(c) = a$. Therefore, $(g \circ f)^{-1}(c) = (f^{-1} \circ g^{-1})(c)$.

Hence, $(g \circ f)^{-1} = (f^{-1} \circ g^{-1})$. ■

Exercises 3.4

1. Prove $f \circ f^{-1} = e_B$ from Theorem 3.4.1, where e_B is the identity function on B.

2. Let f: $A \to B$ such that f is a 1–1 correspondence. Then by Theorem 3.4.2, f^{-1} exists where $f^{-1}: B \to A$ and f^{-1} is a 1–1 correspondence. Hence, $(f^{-1})^{-1}$ exists. Prove $(f^{-1})^{-1} = f$.

3. Let f: $\mathbf{R} \to \mathbf{R}$ where $f(x) = 3x + 5$. Since f is a 1–1 correspondence (see Exercises 3.2, problem (2)), f^{-1} exists. Find $f^{-1}(x)$.

4. Let g: $\mathbf{R} \to \mathbf{R}$ where $g(x) = x^3$. Show g is a 1–1 correspondence and find $g^{-1}(x)$.

5. Using the results of problems (3) and (4) above, find $f \circ g$, $(f \circ g)^{-1}$ and $g^{-1} \circ f^{-1}$. Then verify the result of Theorem 3.4.3, i.e., that $(f \circ g)^{-1} = g^{-1} \circ f^{-1}$.

Section 3.5
FUNCTIONS AND SETS

If f: $A \rightarrow B$, then for each subset of A there corresponds a subset of B and, conversely, each subset of B can be associated with a subset of A. In this section we introduce the concepts of *image of a subset of* A and *inverse image of a subset of* B together with some properties which hold relative to unions, intersections, differences and complements.

Definition 3.5.1

Let f: $A \rightarrow B$.

a. If $X \subseteq A$, **f(X)** denotes the subset of B containing all elements f(x) for each $x \in X$. The set f(X) is called the ***image of X under f.*** Symbolically,

$$f(X) = \{f(x) \in B \mid x \in X\}.$$

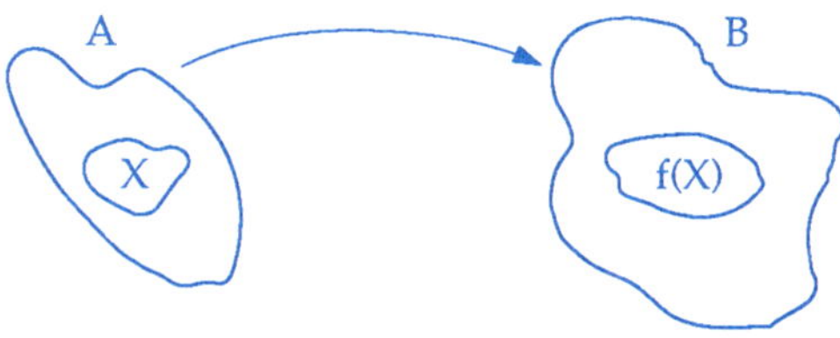

b. If $Y \subseteq B$, $\mathbf{f^{-1}(Y)}$ denotes the subset of A containing all elements $a \in A$ for which $f(a) \in Y$. The set $f^{-1}(Y)$ is called the ***inverse image of Y under f.*** Symbolically,

$$f^{-1}(Y) = \{a \in A \mid f(a) \in Y\}.$$

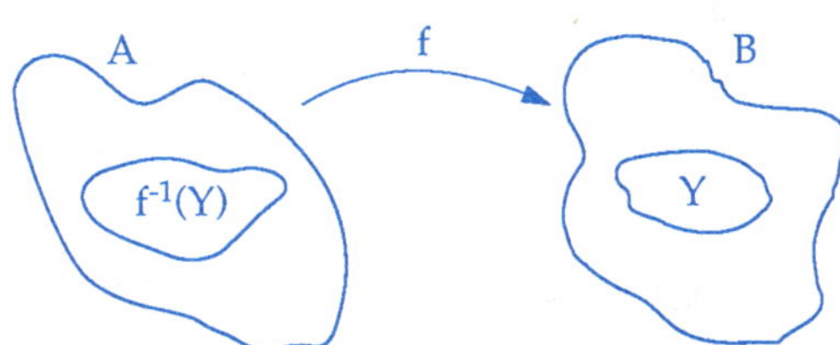

Let f: A ⟶ B with X ⊆ A and Y ⊆ B. Note that f(X) and $f^{-1}(Y)$ are sets. If b ∈ f(X), then b can be written in the form f(x) for some x ∈ X; i.e., b = f(x). If a ∈ $f^{-1}(Y)$, then the image of a under f is an element of Y; i.e., f(a) ∈ Y. Summarizing,

$$b \in f(X) \Leftrightarrow \exists\, x \in X \text{ such that } b = f(x)$$

and

$$a \in f^{-1}(Y) \Leftrightarrow f(a) \in Y.$$

For Y ⊆ B, the notation $f^{-1}(Y)$ does *not* imply that f^{-1} is a function from B to A since f may not be a 1–1 correspondence. If b ∈ B, then {b} ⊆ B and we have to be careful not to confuse the set $f^{-1}(\{b\})$ with the element $f^{-1}(b)$. $f^{-1}(\{b\})$ is the set of all those elements of A whose image is b — it is a subset of A which may consist of more than one element. If f is a 1–1 correspondence, then f^{-1} is a function from B to A and we may speak of the element $f^{-1}(b)$ which is an unique element of A.

Example 3.5.1

a. Let A = {x,y,z,w}, and B = {1,2,3,4,5}. Consider the function f: A ⟶ B where

f: A	⟶	B
x		1
y		2
z		3
w		4
		5

i. If X ⊆ A, then f(X) is the image set of X. That is, f(X) consists only of those elements of B that are images of the elements of X under f.

For X = A ⊆ A, $f(A) = \{f(a) \in B \mid a \in A\}$
$= \{f(x),f(y),f(z),f(w)\}$
$= \{2,4,3\}$
$= R_f$.

For X = {x,y} ⊆ A, $f(\{x,y\}) = \{f(a) \in B \mid a \in \{x,y\}\}$
$= \{f(x),f(y)\}$
$= \{2,4\}$.

For X = {x,z,w} ⊆ A, $f(\{x,z,w\}) = \{f(a) \in B \mid a \in \{x,z,w\}\}$
$= \{f(x),f(z),f(w)\}$
$= \{2,4,3\}$.

For X = ∅ ⊆ A, f(∅) = ∅.

ii. If $Y \subseteq B$, then $f^{-1}(Y)$ is the set of all those elements of A whose image under f is an element of the set Y.

$$\begin{aligned} \text{For } Y = B \subseteq B,\ f^{-1}(B) &= \{a \in A \mid f(a) \in B\} \\ &= \{x,y,z,w\} \\ &= D_f. \end{aligned}$$

$$\begin{aligned} \text{For } Y = \{4\} \subseteq B,\ f^{-1}(\{4\}) &= \{a \in A \mid f(a) \in \{4\}\} \\ &= \{y,z\}. \end{aligned}$$

$$\begin{aligned} \text{For } Y = \{2,3\} \subseteq B,\ f^{-1}(\{2,3\}) &= \{a \in A \mid f(a) \in \{2,3\}\} \\ &= \{x,w\}. \end{aligned}$$

$$\begin{aligned} \text{For } Y = \{1,5\} \subseteq B,\ f^{-1}(\{1,5\}) &= \{a \in A \mid f(a) \in \{1,5\}\} \\ &= \emptyset. \end{aligned}$$

b. Let $f: \mathbf{R} \to \mathbf{R}$ where $f(x) = x^2$.

$$\begin{aligned} f([-2,-1]) &= \{f(x) \in \mathbf{R} \mid -2 \leq x \leq -1\} \\ &= [1,4] \end{aligned}$$

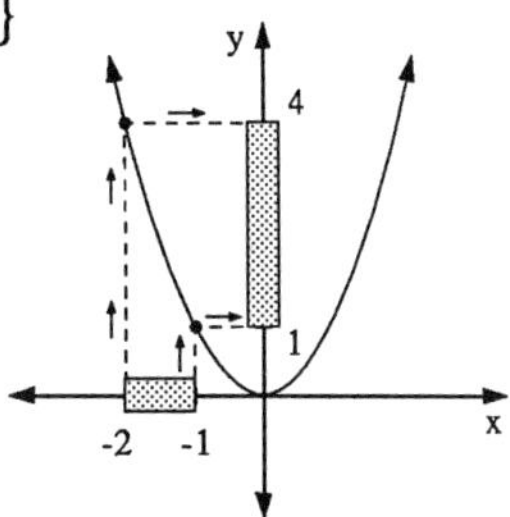

$$\begin{aligned} f^{-1}([1,4]) &= \{x \in \mathbf{R} \mid 1 \leq f(x) \leq 4\} \\ &= [-2,-1] \cup [1,2] \end{aligned}$$

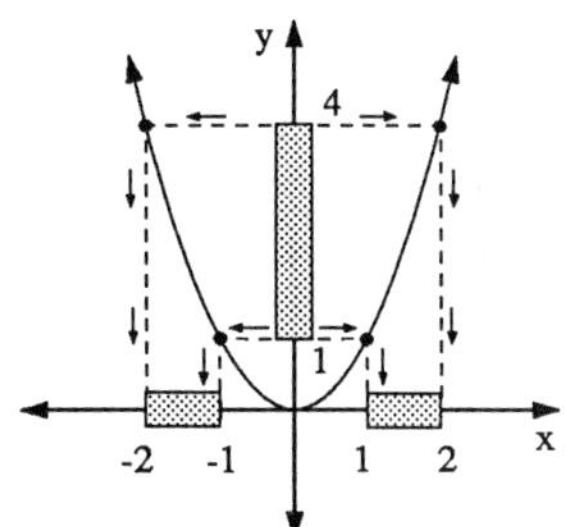

In like manner, we could show that:

$f([-2,2]) = \{f(x) \in \mathbf{R} \mid -2 \leq x \leq 2\} = [0,4]$

$f^{-1}([0,1]) = \{x \in \mathbf{R} \mid 0 \leq f(x) \leq 1\} = [-1,1]$

$f^{-1}([-4,1]) = \{x \in \mathbf{R} \mid -4 \leq f(x) \leq 1\} = [-1,1]$

Theorem 3.5.1

Let f: A $\longrightarrow$ B.

a. $f(\emptyset) = \emptyset$.

b. If $X \subseteq Z \subseteq A$, then $f(X) \subseteq f(Z)$.

c. If $Y \subseteq W \subseteq B$, then $f^{-1}(Y) \subseteq f^{-1}(W)$.

Proof

We shall prove part (b) and leave the proofs of parts (a) and (c) as exercises.

b. Let f: A $\longrightarrow$ B and $X \subseteq Z \subseteq A$. We wish to show $f(X) \subseteq f(Z)$. Let $y \in f(X)$. Then $y = f(x)$ for at least one $x \in X$. But $x \in X \Rightarrow x \in Z$ since $X \subseteq Z$. Therefore $f(x) \in f(Z)$ or $y \in f(Z)$. Hence, $f(X) \subseteq f(Z)$.

■

Let f: A $\longrightarrow$ B. Theorem 3.5.1 shows that when moving subsets from A to B or B to A set containment ($\subseteq$) is always preserved. If f: A $\longrightarrow$ B is either not 1–1 or not onto, moving a subset back and forth between A and B may cause you to gain or lose elements depending upon what properties the function f possesses as the following example exhibits. The following example, as well as Example 3.5.1, part (b), exhibit this difficulty.

Example 3.5.2

Let A = {a,b,c,d} and B = {1,2,3,4}. Consider the function f: A $\longrightarrow$ B where

f: A $\longrightarrow$ B

a → 2
b → 1
c → 2
d → 3

(B = 1, 2, 3, 4)

When X = {a,b}, f(X) = {1,2} and $f^{-1}(f(X))$ = {a,b,c}.
Hence, $X \subseteq f^{-1}(f(X))$, but $f^{-1}(f(X) \neq X$.

When Y = {3,4}, $f^{-1}(Y)$ = {d} and $f(f^{-1}(Y))$ = {3}.
Hence, $f(f^{-1}(Y)) \subseteq Y$, but $f(f^{-1}(Y)) \neq Y$.

Theorem 3.5.2

Let $f: A \to B$, $X \subseteq A$ and $Y \subseteq B$. Then,

a. $f(f^{-1}(Y)) \subseteq Y$.

b. If f is onto, then $f(f^{-1}(Y)) = Y$.

c. $X \subseteq f^{-1}(f(X))$.

d. If f is 1–1, then $f^{-1}(f(X)) = X$.

Proof

We shall prove parts (a) and (d) and leave the proofs of parts (b) and (c) as exercises.

a. Let $y \in f(f^{-1}(Y))$. Hence, $\exists a \in f^{-1}(Y)$ such that $y = f(a)$. Since $a \in f^{-1}(Y)$, we have $f(a) \in Y$ by Definition 3.5.1. Therefore, $y \in Y$ and $f(f^{-1}(Y)) \subseteq Y$. ■

d. Let f be 1–1. We wish to show $X = f^{-1}(f(X))$; i.e., $X \subseteq f^{-1}(f(X))$ and $f^{-1}(f(X)) \subseteq X$. By part (c), $X \subseteq f^{-1}(f(X))$. Hence, to complete the proof, we need only show $f^{-1}(f(X)) \subseteq X$.

Let $a \in f^{-1}(f(X))$. Then $f(a) \in f(X)$. Since $f(a) \in f(X)$, then $f(a)$ can be written in the form $f(x)$ for some $x \in X$. That is, $f(a) = f(x)$. But f is 1–1. Hence, $a = x$. Therefore, since $x \in X$, $a \in X$ and we have $f^{-1}(f(X)) \subseteq X$. ■

The next two theorems together with problems (5) and (6) in Exercises 3.5 show those properties governing unions, intersections, differences and complements for images and inverse images of subsets.

Theorem 3.5.3

Let $f: A \to B$, and $X, Z \subseteq A$. Then,

a. $f(X \cup Z) = f(X) \cup f(Z)$.

b. $f(X \cap Z) \subseteq f(X) \cap f(Z)$.

Proof

a. Let $y \in f(X \cup Z)$. Then $\exists x \in X \cup Z$ such that $f(x) = y$. Now $x \in X \cup Z$ implies $x \in X$ or $x \in Z$. Therefore, $f(x) \in f(X)$ or $f(x) \in f(Z)$. Hence, $f(x) \in f(X) \cup f(Z)$. Thus $y \in f(X) \cup f(Z)$ and $f(X \cup Z) \subseteq f(X) \cup f(Z)$.

Let $y \in f(X) \cup f(Z)$. Then $y \in f(X)$ or $y \in f(Z)$. So $\exists x \in X$ such that $f(x) = y$ or $\exists z \in Z$ such that $f(z) = y$. Therefore $x \in X \cup Z$ or $z \in X \cup Z$. Hence, $f(x) \in f(X \cup Z)$ or $f(z) \in f(X \cup Z)$. Thus $y \in f(X \cup Z)$ and $f(X) \cup f(Z) \subseteq f(X \cup Z)$.

Therefore, $f(X \cup Z) = f(X) \cup f(Z)$. ■

b. Let $y \in f(X \cap Z)$. Then $\exists x \in X \cap Z$ such that $f(x) = y$. Now $x \in X \cap Z$ implies $x \in X$ and $x \in Z$. Therefore, $f(x) \in f(X)$ and $f(x) \in f(Z)$. Hence, $f(x) \in f(X) \cap f(Z)$. Thus $y \in f(X) \cap f(Z)$ and $f(X \cap Z) \subseteq f(X) \cap f(Z)$.

■

Example 3.5.3

Let $A = \{x,y,z\}$ and $B = \{1,2,3,4\}$. Consider the function $f: A \to B$ where

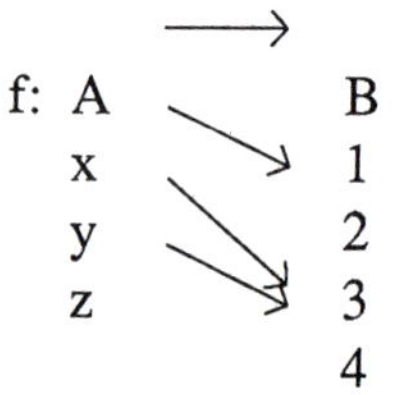

f: A ⟶ B
x 1
y 2
z 3
4

$X = \{x,y\}$	$f(X) = \{2,4\}$
$Z = \{x,z)$	$f(Z) = \{2,4\}$
$X \cap Z = \{x\}$	$f(X \cap Z) = \{2\}$ and $f(X) \cap f(Z) = \{2,4\}$

Hence, $f(X \cap Z) \subseteq f(X) \cap f(Z)$ but $f(X \cap Z) \neq f(X) \cap f(Z)$.

Theorem 3.5.4

Let $f: A \to B$, and $Y, W \subseteq B$. Then,

a. $f^{-1}(Y \cup W) = f^{-1}(Y) \cup f^{-1}(W)$.

b. $f^{-1}(Y \cap W) = f^{-1}(Y) \cap f^{-1}(W)$.

Proof

Let f: A → B and Y, W ⊆ B. We shall prove part (a) and leave the proof of part (b) as an exercise.

a. We must show $f^{-1}(Y \cup W) = f^{-1}(Y) \cup f^{-1}(W)$.

$$\begin{aligned} x \in f^{-1}(Y \cup W) &\Rightarrow f(x) \in Y \cup W \\ &\Rightarrow f(x) \in Y \text{ or } f(x) \in W \\ &\Rightarrow x \in f^{-1}(Y) \text{ or } x \in f^{-1}(W) \\ &\Rightarrow x \in f^{-1}(Y) \cup f^{-1}(W) \end{aligned}$$

Hence, $f^{-1}(Y \cup W) \subseteq f^{-1}(Y) \cup f^{-1}(W)$.

$$\begin{aligned} x \in f^{-1}(Y) \cup f^{-1}(W) &\Rightarrow x \in f^{-1}(Y) \text{ or } x \in f^{-1}(W) \\ &\Rightarrow f(x) \in Y \text{ or } f(x) \in W \\ &\Rightarrow f(x) \in Y \cup W \\ &\Rightarrow x \in f^{-1}(Y \cup W) \end{aligned}$$

Hence, $f^{-1}(Y) \cup f^{-1}(W) \subseteq f^{-1}(Y \cup W)$

Therefore, $f^{-1}(Y \cup W) = f^{-1}(Y) \cup f^{-1}(W)$. ■

Example 3.5.4

Let A = {a,b,c,d} and B = {1,2,3,4,5}. Consider the function f: A → B where

f: A ⟶ B
a → 2
b → 1
c → 4
d → 5

(3 has no preimage)

Y = {2,3,5}	$f^{-1}(Y) = \{a,d\}$
W = {3,4,5}	$f^{-1}(W) = \{c,d\}$
$Y \cup W = \{2,3,4,5\}$	$f^{-1}(Y \cup W) = \{a,c,d\} = f^{-1}(Y) \cup f^{-1}(W)$
$Y \cap W = \{3,5\}$	$f^{-1}(Y \cap W) = \{d\} = f^{-1}(Y) \cap f^{-1}(W)$

Exercises 3.5

1. Let $A = \{a,b,c,d,e\}$ and $B = \{1,2,3,4,5,6\}$. Consider the function $f: A \to B$ where

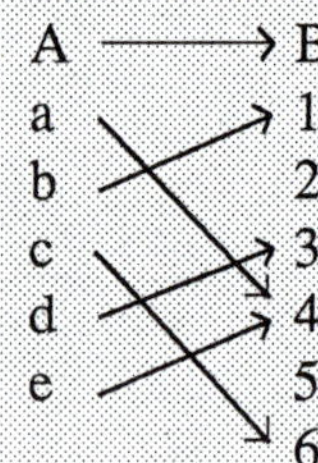

   a. Let $X \subseteq A$ where $X = \{a,c,d\}$. Find $f(X)$ and $f^{-1}(f(X))$.

   b. Let $Y \subseteq B$ where $Y = \{1,2,4,5\}$. Find $f^{-1}(Y)$ and $f(f^{-1}(Y))$.

2. Prove parts (a) and (c) of Theorem 3.5.1.

3. Let $f: A \to B$ and $Y \subseteq B$. Prove $f(f^{-1}(Y)) = f(A) \cap Y$.

4. Prove parts (b) and (c) of Theorem 3.5.2.

5. Let $f: A \to B$ and $X, Z \subseteq A$. If f is 1–1, prove $f(X \cap Z) = f(X) \cap f(Z)$.

6. Let $f: A \to B$.

   a. If $Y, W \subseteq B$, prove $f^{-1}(Y - W) = f^{-1}(Y) - f^{-1}(W)$.

   b. If $X, Z, \subseteq A$, show by means of an example that $f(X - Z) \neq f(X) - f(Z)$.

   c. If $X, Z \subseteq A$, prove $f(X) - f(Z) \subseteq f(X - Z)$

   d. If $X, Z \subseteq A$ and f is 1–1, prove $f(X - Z) = f(X) - f(Z)$.

7. Let $f: A \to B$ and $Y \subseteq B$. Prove $f^{-1}(Y') = [f^{-1}(Y)]'$ where $Y' = B - Y$.

8. Prove part (b) of Theorem 3.5.4.

Chapter 4

RELATIONS AND PARTITIONS

In Chapter 2, Section 2.5, we discussed the idea of an ordered pair of elements (a,b) together with the product set A x B where

$$A \times B = \{(a,b) \mid a \in A \wedge b \in B\}.$$

In this chapter we shall utilize these ideas to introduce the concept of a *mathematical relation*. Of particular interest will be the mathematical relations referred to as *equivalence relations* — relations which possess properties quite similar to those of the relation of equality. The final concept discussed in this chapter is that of a *partition* of a set. We shall show the existence of a very important connection between an equivalence relation on a set and a partition of that set. The notion of a relation will also enable us to develop an alternate definition of a function which may be thought of as a particular type of relation.

Section 4.1
RELATIONS

You are familiar with the word "relation" as it is used in everyday language. A person is related to their mother, father, sister or brother. The word "relation" may also be extended to such situations as: two persons are related in the sense of being neighbors, in the sense of living in the same county or in the sense of being classmates (freshmen, sophomores, juniors or seniors) at a particular university. The mathematical usage of *relation* is very similar and is also based on the concept of a pair.

Definition 4.1.1

Let A and B be two sets. A ***relation*** from A to B, denoted by the symbol $\mathcal{R}$, is any subset of A x B.

Example 4.1.1

Let A ={a,b,c} and B = {1,2}. Then

$$A \times B = \{(a,1),(a,2),(b,1),(b,2),(c,1),(c,2)\}.$$

Now any subset of A x B (note that there are 2^6 subsets of A x B) is called a relation from A to B. Listed below are three such relations.

a. $\mathcal{R}_1 = \{(a,1),(b,2),(c,1)\}$

b. $\mathcal{R}_2 = \{(b,1),(c,2)\}$

c. $\mathcal{R}_3 = \{(a,1),(b,2),(c,1),(c,2)\}$

Let $\mathcal{R}$ be a relation from A to B. If $(a,b) \in \mathcal{R}$, we say "*a is related to b*" and usually write $a \mathcal{R} b$. Hence, $(a,b) \in \mathcal{R}$ and $a \mathcal{R} b$ have the same meaning. In like manner, if $(a,b) \notin \mathcal{R}$, we say "*a is not related to b*" and write $a \not\mathcal{R} b$ as an equivalent statement for $(a,b) \notin \mathcal{R}$. For instance, refer to Example 4.1.1, part (a), where $\mathcal{R}_1 = \{(a,1),(b,2),(c,1)\}$. Since $(b,2) \in \mathcal{R}_1$, $b \mathcal{R}_1 2$. However, $a \not\mathcal{R}_1 2$ since $(a,2) \notin \mathcal{R}_1$.

The terms domain, range and inverse had specific meanings when we discussed functions in the previous chapter. They have analogous meanings with regard to relations. The use of these analogous meanings will help us to develop an alternate definition of a function.

Definition 4.1.2

Let $\mathcal{R}$ be a relation from A to B.

a. The ***domain of*** $\mathcal{R}$, denoted by $D_{\mathcal{R}}$, is the set of all the first elements of the ordered pairs belonging to $\mathcal{R}$. Symbolically,

$$D_{\mathcal{R}} = \{a \in A \mid (a,b) \in \mathcal{R}\}.$$

b. The ***range of*** $\mathcal{R}$, denoted by $R_{\mathcal{R}}$, is the set of all the second elements of the ordered pairs belonging to $\mathcal{R}$. Symbolically,

$$R_{\mathcal{R}} = \{b \in B \mid (a,b) \in \mathcal{R}\}.$$

Example 4.1.2

Let $A = \{a,b,c\}$ and $B = \{1,2\}$.

a. If $\mathcal{R}_1 = \{(a,1),(b,2),(c,1)\}$, then $D_{\mathcal{R}_1} = \{a,b,c\}$ and $R_{\mathcal{R}_1} = \{1,2\}$.

b. If $\mathcal{R}_2 = \{(b,1), (c,2)\}$, then $D_{\mathcal{R}_2} = \{b,c\}$ and $R_{\mathcal{R}_2} = \{1,2\}$.

Definition 4.1.3

Let A an B be two sets and $\mathcal{R}$ a relation from A to B. Then the ***inverse*** of the relation $\mathcal{R}$, denoted by $\mathcal{R}^{-1}$, is a relation from B to A such that if $(a,b) \in \mathcal{R}$, then $(b,a) \in \mathcal{R}^{-1}$. Symbolically,

$$\mathcal{R}^{-1} = \{(b,a) \mid (a,b) \in \mathcal{R}\}.$$

Example 4.1.3

Let A ={a,b,c} and B = {1,2}.

a. If $\mathcal{R}_1 = \{(a,1),(b,2),(c,1)\}$, then $\mathcal{R}_1^{-1} = \{(1,a),(2,b),(1,c)\}$

b. If $\mathcal{R}_2 = \{(b,1),(c,2)\}$, then $\mathcal{R}_2^{-1} = \{(1,b),(2,c)\}$

c. If $\mathcal{R}_3 = \{(a,1),(b,2),(c,1),(c,2)\}$, then $\mathcal{R}_3^{-1} = \{(1,a),(2,b),(1,c),(2,c)\}$.

The notions of a relation from A to B, its domain and its range afford us an alternative approach to the concept of a function from set A to set B.

Definition 4.1.4

Let A and B be two non-empty sets. A ***function*** from A to B is a relation, $\mathcal{R}$, from A to B such that:

a. $D_{\mathcal{R}} = A$.

b. If $(a,b) \in \mathcal{R}$ and $(a,c) \in \mathcal{R}$, then $b = c$.

Note that condition (a) of the definition assures that each element of A corresponds to at least one element of set B while condition (b) assures that each element of A corresponds to one and only one element of set B. If the relation $\mathcal{R}$ is a function from A to B, we shall denote $\mathcal{R}$ by f, g, etc., to conform with our previous notation.

Example 4.1.4

Let A = {a,b,c} and B = {1,2}.

a. If $\mathcal{R}_1 = \{(a,1),(b,2),(c,1)\}$, then $\mathcal{R}_1$ is a function from A to B.

b. If $\mathcal{R}_2 = \{(b,1),(c,2)\}$, then $\mathcal{R}_2$ is not a function from A to B since $D_{\mathcal{R}_2} = \{b,c\} \neq A$.

c. If $\mathcal{R}_3 = \{(a,1),(b,2),(c,1),(c,2)\}$, then $\mathcal{R}_3$ is not a function from A to B since $(c,1) \in \mathcal{R}_3$ and $(c,2) \in \mathcal{R}_3$ but $1 \neq 2$.

Let f be a function from A to B. Then if $(a,b) \in f$, it is customary to write $b = f(a)$. The concepts and definitions developed in Chapter 3 regarding functions thus could all be re-written in terms of this alternate definition of a function. For example, f is **onto** if for each $b \in B$ there exists at least one $a \in A$ such that $(a,b) \in f$. We shall leave the other concepts and corresponding definitions as exercises.

Exercises 4.1

1. Let $A = \{a,b\}$ and $B = \{1,2\}$. Find all relations from A to B. (Label the relations $\mathcal{R}_1$, $\mathcal{R}_2$, etc.)

2. For each relation $\mathcal{R}_i$, found in problem (1), find $\mathcal{R}_i^{-1}$.

3. Which relations $\mathcal{R}_i$, found in problem (1) are functions from A to B?

4. Using the results of problem (3), which inverse relations, $\mathcal{R}_i^{-1}$, from B to A are functions from B to A? (If $\mathcal{R}_i$ is a function from A to B, what properties must $\mathcal{R}_i$ possess in order for $\mathcal{R}_i^{-1}$ to be a function from B to A?)

5. Rewrite the following definitions in terms of Definition 4.1.4 for a function $f: A \to B$.
   a. **domain** of f
   b. **range** of f
   c. f is **1–1**
   d. the **inverse function** of f, if it exists.
   e. if $g: B \to C$, $g \circ f$

Section 4.2
RELATIONS ON A FIXED SET

Of particular importance are those relations from a set to itself. In this way elements from a given set are related to other elements within that same set. For instance, in the set of real numbers you are familiar with the relations "is equal to" and "is less than." In this section we shall formalize the concept involved in these relations.

Definition 4.2.1

A ***relation*** on a set A is a relation, $\mathcal{R}$, from A to A; i.e., $\mathcal{R}$ is any subset of A x A.

Example 4.2.1

Let A = {a,b,c}. Then

A x A = {(a,a),(a,b),(a,c),(b,a),(b,b),(b,c),(c,a),(c,b),(c,c)}.

Now any subset of A x A is called a relation on A. Each of the following sets is a relation on A:

a. $\mathcal{R}_1$ = {(a,a),(b,b),(c,c),(a,b)}.

b. $\mathcal{R}_2$ = {(a,a),(b,b),(c,c),(a,b),(b,a),(a,c),(c,a)}.

c. $\mathcal{R}_3$ = {(a,a),(b,b),(a,b),(b,a)}.

Example 4.2.2

Each of the following is a relation on **R**, the set of real numbers.

a. For x, y ∈ **R**, $x \mathcal{R}_1 y \Leftrightarrow x \le y$;
i.e., $\mathcal{R}_1 = \{(x,y) \in \mathbf{R} \times \mathbf{R} \mid x \le y\}$.

Hence $\mathcal{R}_1$ consists of all (x,y) ∈ **R** x **R** which lie on or above the line y = x.

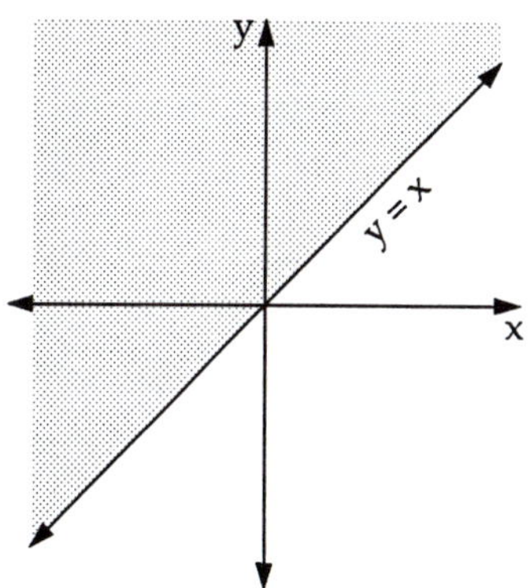

b. For x, y ∈ **R**, $x \mathcal{R}_2 y \Leftrightarrow 2x - 3y = 6$;
i.e., $\mathcal{R}_2 = \{(x,y) \in \mathbf{R} \times \mathbf{R} \mid 2x - 3y = 6\}$.

Hence, $\mathcal{R}_2$ consists of all (x,y) ∈ **R** x **R** which lie on the line 2x – 3y = 6.

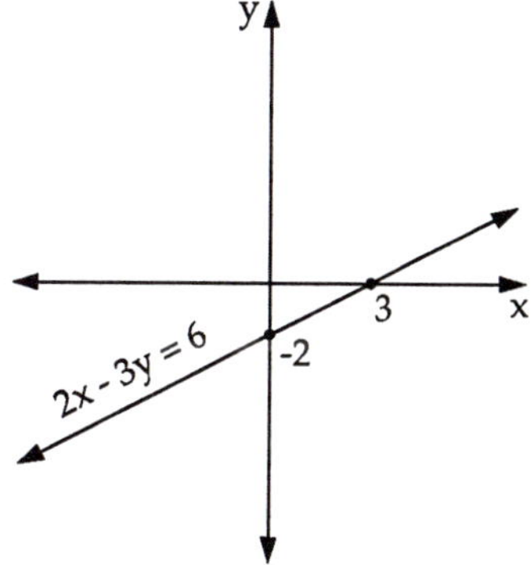

c. For x, y ∈ **R**, $x \mathcal{R}_3 y \Leftrightarrow x + y < 1$;
i.e., $\mathcal{R}_3 = \{(x,y) \in \mathbf{R} \times \mathbf{R} \mid x + y < 1\}$.

Hence, $\mathcal{R}_3$ consists of all (x,y) ∈ **R** x **R** which lie below the line x + y = 1.

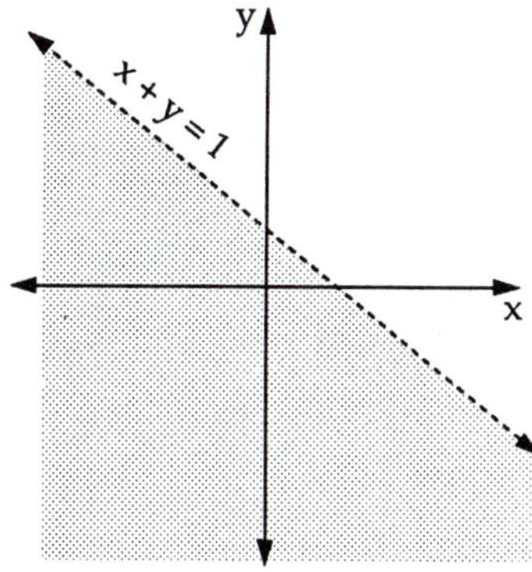

Example 4.2.3

Let $\mathcal{L}$ be the set of all straight lines in the Euclidean plane. Each of the following is a relation on $\mathcal{L}$:

a. For L, K $\in \mathcal{L}$, L $\mathcal{R}_1$K $\Leftrightarrow$ L is parallel to K; i.e., lines L and K have the same slope ($m_L = m_K$). Hence, $\mathcal{R}_1 = \{(L,K) \in \mathcal{L} \times \mathcal{L} \mid L \parallel K\}$.

b. For L, K $\in \mathcal{L}$, L $\mathcal{R}_2$K $\Leftrightarrow$ L is perpendicular to K; i.e., lines L and K have slopes that are the negative reciprocals of each other ($m_L = -\frac{1}{m_K}$). Hence, $\mathcal{R}_2 = \{(L,K) \in \mathcal{L} \times \mathcal{L} \mid L \perp K\}$.

Example 4.2.4

Let $X \neq \emptyset$ and P(X) the power set of X. For A, B $\in$ P(X), let A $\mathcal{R}$ B $\Leftrightarrow$ A $\subseteq$ B. Then $\mathcal{R}$ is a relation on P(X); i.e., $\mathcal{R} \subseteq P(X) \times P(X)$. Hence, if we let X = {a,b}, then P(X) = $\{\emptyset,\{a\},\{b\},X\}$. Recall that $\emptyset \subseteq$ A and A $\subseteq$ A for any set A. Thus the relation $\mathcal{R}$ on P(X) consists of the elements:

$$\mathcal{R} = \{(\emptyset,\emptyset),(\emptyset,\{a\}),(\emptyset,\{b\}),(\emptyset,X),(\{a\},\{a\}),(\{a\},X),(\{b\},\{b\}),(\{b\},X),(X,X)\}$$

Exercises 4.2

Let A = {1,2}. Answer the following:

1. Find all relations on A. (Label the relations $\mathcal{R}_1$, $\mathcal{R}_2$, etc.)

2. Which relations $\mathcal{R}_i$, found in problem (1) are functions from A to A?

3. Using the results of problem (2), which of the relations have inverses that are also functions from A to A?

Section 4.3
TYPES OF RELATIONS ON A FIXED SET

A relation $\mathcal{R}$ on a given set may be classified according to the properties which that relation possesses. We wish to consider three classifications of a relation — *reflexive*, *symmetric* and *transitive*.

Definition 4.3.1

Let $\mathcal{R}$ be a relation on A.

a. $\mathcal{R}$ is ***reflexive*** if, $\forall a \in A$, $a \mathcal{R} a$.

$$(\forall a \in A, (a,a) \in \mathcal{R})$$

b. $\mathcal{R}$ is ***symmetric*** if, whenever $a \mathcal{R} b$, then $b \mathcal{R} a$.

$$((a,b) \in \mathcal{R} \Rightarrow (b,a) \in \mathcal{R})$$

c. $\mathcal{R}$ is ***transitive*** if, whenever $a \mathcal{R} b$ and $b \mathcal{R} c$, then $a \mathcal{R} c$.

$$((a,b) \in \mathcal{R} \text{ and } (b,c) \in \mathcal{R} \Rightarrow (a,c) \in \mathcal{R})$$

Example 4.3.1

a. Refer to Example 4.2.1, where $A = \{a,b,c\}$.

$\mathcal{R}_1 = \{(a,a),(b,b),(c,c),(a,b)\}$ is reflexive and transitive. $\mathcal{R}_1$ is not symmetric since $a \mathcal{R}_1 b$, but $b \not\mathcal{R}_1 a$ $((a,b) \in \mathcal{R}_1$, but $(b,a) \notin \mathcal{R}_1)$.

$\mathcal{R}_2 = \{(a,a),(b,b),(c,c),(a,b),(b,a),(a,c),(c,a)\}$ is reflexive and symmetric. $\mathcal{R}_2$ is not transitive since $b \mathcal{R}_2 a$ and $a \mathcal{R}_2 c$, but $b \not\mathcal{R}_2 c$ $((b,a) \in \mathcal{R}_2$ and $(a,c) \in \mathcal{R}_2$, but $(b,c) \notin \mathcal{R}_2)$.

$\mathcal{R}_3 = \{(a,a),(b,b),(a,b),(b,a)\}$ is symmetric and transitive. $\mathcal{R}_3$ is not reflexive since $c \in A$, but $c \not\mathcal{R}_3 c$ $(c \in A$, but $(c,c) \notin \mathcal{R}_3)$.

Example 4.3.1 continued next page.

Example 4.3.1 Continued

b. Refer to Example 4.2.3, part (b).

For L, K $\in \mathcal{L}$, L $\mathcal{R}_2$ K $\Leftrightarrow$ L is perpendicular to K. This relation is symmetric. However, $\mathcal{R}_2$ is not reflexive since each line L $\in \mathcal{L}$ is parallel to itself. Also, $\mathcal{R}_2$ is not transitive since, if L $\mathcal{R}_2$ K and K $\mathcal{R}_2$ J, then L and J are parallel. Hence, L $\not\mathcal{R}_2$ J.

c. Refer to Example 4.2.4. For A, B $\in$ P(X), A $\mathcal{R}$ B $\Leftrightarrow$ A $\subseteq$ B.

This relation is reflexive and transitive. However, $\mathcal{R}$ is not symmetric. From our example where X = {a,b}, we see that if A $\subseteq$ B, it is not necessarily true that B $\subseteq$ A. Hence, A $\mathcal{R}$ B does not imply B $\mathcal{R}$ A.

Relations that possess all three of these properties — reflexive, symmetric and transitive — are extremely important and are referred to as *equivalence relations*. In the next section we shall discuss in detail these particular relations and how useful they are in analyzing the elements of a set.

Definition 4.3.2

Let $\mathcal{R}$ be a relation on A. $\mathcal{R}$ is called an ***equivalence relation*** on A if $\mathcal{R}$ is reflexive, symmetric and transitive.

Hence, if one wishes to show a relation, $\mathcal{R}$, on A is an equivalence relation, one must demonstrate that $\mathcal{R}$ possesses each of the three properties: reflexive, symmetric and transitive.

Example 4.3.2

a. Refer to Example 4.2.3, part (a). For L, K $\in \mathcal{L}$, L $\mathcal{R}_1$ K $\Leftrightarrow$ L is parallel to K. $\mathcal{R}_1$ is an equivalence relation on $\mathcal{L}$.

b. Let A = {a,b,c,d}. Verify $\mathcal{R}$ = {(a,a),(b,b),(c,c),(d,d),(b,d),(d,b)} is an equivalence relation on A.

We must show $\forall x \in A, (x,x) \in \mathcal{R}$:

$a \in A$,	$(a,a) \in \mathcal{R}$
$b \in A$,	$(b,b) \in \mathcal{R}$
$c \in A$,	$(c,c) \in \mathcal{R}$
$d \in A$,	$(d,d) \in \mathcal{R}$

Hence, $\mathcal{R}$ is reflexive.

We must show if $(x,y) \in \mathcal{R}$, then $(y,x) \in \mathcal{R}$.

$(a,a) \in \mathcal{R}$,	$(a,a) \in \mathcal{R}$
$(b,b) \in \mathcal{R}$,	$(b,b) \in \mathcal{R}$
$(c,c) \in \mathcal{R}$,	$(c,c) \in \mathcal{R}$
$(d,d) \in \mathcal{R}$,	$(d,d) \in \mathcal{R}$
$(b,d) \in \mathcal{R}$,	$(d,b) \in \mathcal{R}$
$(d,b) \in \mathcal{R}$,	$(b,d) \in \mathcal{R}$

Hence, $\mathcal{R}$ is symmetric.

We must show if $(x,y) \in \mathcal{R}$ and $(y,z) \in \mathcal{R}$, then $(x,z) \in \mathcal{R}$:

$(b,b), (b,d) \in \mathcal{R}$,	$(b,d) \in \mathcal{R}$
$(d,d), (d,b) \in \mathcal{R}$,	$(d,b) \in \mathcal{R}$
$(b,d), (d,d) \in \mathcal{R}$,	$(b,d) \in \mathcal{R}$
$(b,d), (d,b) \in \mathcal{R}$,	$(b,b) \in \mathcal{R}$
$(d,b), (b,b) \in \mathcal{R}$,	$(d,b) \in \mathcal{R}$
$(d,b), (b,d) \in \mathcal{R}$,	$(d,d) \in \mathcal{R}$

Hence, $\mathcal{R}$ is transitive.

Since $\mathcal{R}$ is reflective, symmetric and transitive, $\mathcal{R}$ is an equivalence relation on A.

Example 4.3.2 continued next page.

Example 4.3.2 Continued

c. Let **Z** be the set of integers. For a, b $\in$ **Z**, let a $\mathcal{R}$ b $\Leftrightarrow$ a – b = 3k, k $\in$ **Z**; i.e., $\mathcal{R}$ = {(a, b) $\in$ **Z** x **Z** | a – b = 3k, k $\in$ **Z**}. We wish to verify that $\mathcal{R}$ is an equivalence relation on **Z**. (Note that 6 $\mathcal{R}$ 12 since 6 – 12 = -6 = 3(-2), while 6 $\not\mathcal{R}$ (-4) since 6 – (-4) = 10 $\neq$ 3k where k $\in$ **Z**.)

For each a $\in$ **Z**, we must show a $\mathcal{R}$ a. Now a – a = 0 = 3(0). Since 0 $\in$ **Z**, a $\mathcal{R}$ a. Hence, $\mathcal{R}$ is reflexive.

Let a $\mathcal{R}$ b. We must show b $\mathcal{R}$ a. Now a $\mathcal{R}$ b $\Rightarrow$ a – b = 3k, k $\in$ **Z**. Therefore -(a – b) = -(3k) $\Rightarrow$ b – a = 3(-k). Since -k $\in$ Z, b $\mathcal{R}$ a. Hence, $\mathcal{R}$ is symmetric.

Let a $\mathcal{R}$ b and b $\mathcal{R}$ c. We must show a $\mathcal{R}$ c. Now a $\mathcal{R}$ b and b $\mathcal{R}$ c $\Rightarrow$ a – b = 3p and b – c = 3q where p, q $\in$ **Z**. Therefore,

$$\begin{aligned} a - c &= (a - b) + (b - c) \\ &= 3p + 3q \\ &= 3(p + q) \\ &= 3k, \text{ where } k = p + q. \end{aligned}$$

Since k $\in$ **Z**, a $\mathcal{R}$ c. Hence, $\mathcal{R}$ is transitive.

Since $\mathcal{R}$ is reflexive, symmetric and transitive, $\mathcal{R}$ is an equivalence relation on **Z**.

d. Let S = {(a,b) | a, b $\in$ Z, b $\neq$ 0}. For (a,b), (c,d) $\in$ S, let (a,b) $\mathcal{R}$ (c,d) $\Leftrightarrow$ ad = bc. We wish to verify that $\mathcal{R}$ is an equivalence relation on S. (Note that for each ordered pair contained in S, the element in the second position must be a non-zero integer.)

For each pair (a,b) $\in$ S, we must show (a,b) $\mathcal{R}$ (a,b). Now ab = ab $\Rightarrow$ ab = ba. Hence, (a,b) $\mathcal{R}$ (a,b) and $\mathcal{R}$ is reflexive.

Let (a,b) $\mathcal{R}$ (c,d). We must show (c,d) $\mathcal{R}$ (a,b). Now (a,b) $\mathcal{R}$ (c,d) $\Rightarrow$ ad = bc. Therefore bc = ad $\Rightarrow$ cb = da. Hence (c,d) $\mathcal{R}$ (a,b) and $\mathcal{R}$ is symmetric.

Let (a,b) $\mathcal{R}$ (c,d) and (c,d) $\mathcal{R}$ (e,f). We must show (a,b) $\mathcal{R}$ (e,f). Now (a,b) $\mathcal{R}$ (c,d) $\Rightarrow$ ad = bc and (c,d) $\mathcal{R}$ (e,f) $\Rightarrow$ cf = de. Thus,

$$
\begin{aligned}
ad &= bc \\
(ad)f &= (bc)f \\
(af)d &= b(cf) \\
(af)d &= b(de) && cf = de \\
(af)d &= (be)d \\
af &= be && d \neq 0.
\end{aligned}
$$

Hence, (a,b) $\mathcal{R}$ (e,f) and $\mathcal{R}$ is transitive.

Since $\mathcal{R}$ is reflexive, symmetric and transitive, $\mathcal{R}$ is an equivalence relation on S.

Exercises 4.3

1. Determine whether or not the given relation, $\mathcal{R}$, is reflexive, symmetric, transitive.

   a. Let A = {a,b,c,d} and $\mathcal{R}$ = {(a,a),(b,b),(d,d),(a,c),(c,a)}.

   b. Let **Z** be the set of integers. For x, y ∈ **Z**, let x $\mathcal{R}$y ⇔ x is a factor of y. (If y = xz, where x, y, z ∈ **Z**, then x and z are called factors of y.)

2. Let A = {1,2,3,4}. Construct a relation $\mathcal{R}$ on A such that:

   a. $\mathcal{R}$ is reflexive but neither symmetric nor transitive.

   b. $\mathcal{R}$ is symmetric but neither reflexive nor transitive.

   c. $\mathcal{R}$ is transitive but neither reflexive nor symmetric.

For 3-6 show each of the following relations, $\mathcal{R}$, defined on the given set is an equivalence relation.

3. Let **R** be the set of real numbers. For x, y ∈ **R**, let x $\mathcal{R}$y ⇔ x − y ∈ **Z**.

4. Let f: A → B where f is fixed. For a, b ∈ A, let a $\mathcal{R}$ b ⇔ f(a) = f(b).

5. Let F = {f | f: A → B} and let a ∈ A, where a is fixed. For f, g ∈ F, let
   f $\mathcal{R}$g ⇔ f(a) = g(a).

6. Let **Z** be the set of integers. For (a,b), (c,d) ∈ **Z** x **Z**, let
   (a,b) $\mathcal{R}$(c,d) ⇔ a + d = b + c.

Section 4.4
EQUIVALENCE SETS

In this section we devote our attention entirely to equivalence relations which occur quite often throughout the various branches of mathematics. They afford us a way of categorizing elements within a given set A as being "equivalent." As we shall see, an equivalence relation will generate a family of subsets of A by specifying certain elements of A as being "equivalent" according to a given rule. Specifically, two elements will be considered "equivalent" if they belong to the same subset generated by the given equivalence relation. Since more than one equivalence relation may exist for a given set A, it should be noted that two elements of A may be viewed as "equivalent" under one relation, but considered not "equivalent" under another relation. This is due to the fact that two different equivalence relations defined on A may not generate the same family of subsets of A.

Definition 4.4.1

Let $\mathcal{R}$ be an equivalence relation on A. For each a ∈ A, the set

$$\overline{a} = \{x \in A \mid x \ \mathcal{R} \ a\} = \{x \in A \mid (x,a) \in \mathcal{R}\}$$

is called an ***equivalence set of*** a determined by the relation $\mathcal{R}$.

Note that an equivalence set determined by a relation $\mathcal{R}$ on A exists only if $\mathcal{R}$ is an equivalence relation on A. Also, note that $x \in \overline{a}$ if and only if $x \ \mathcal{R} \ a$.

Example 4.4.1

Refer to Example 4.3.2.

a. In part (b), where $A = \{a,b,c,d\}$ and $\mathcal{R} = \{(a,a),(b,b),(c,c),(d,d),(b,d),(d,b)\}$, we have

$$\overline{a} = \{x \in A \mid x \ \mathcal{R} \ a\} = \{x \in A \mid (x, a) \in \mathcal{R}\} = \{a\},$$

$$\overline{b} = \{x \in A \mid x \ \mathcal{R} \ b\} = \{x \in A \mid (x, b) \in \mathcal{R}\} = \{b,d\},$$

$$\overline{c} = \{x \in A \mid x \ \mathcal{R} \ c\} = \{x \in A \mid (x, c) \in \mathcal{R}\} = \{c\}.$$

$$\overline{d} = \{x \in A \mid x \ \mathcal{R} \ d\} = \{x \in A \mid (x, d) \in \mathcal{R}\} = \{d,b\}.$$

b. In part (c), where for a, b $\in$ Z, a $\mathcal{R}$ b $\Leftrightarrow$ a − b = 3k, k $\in$ Z, we have for each a $\in$ Z,

$$\overline{a} = \{x \in Z \mid x \,\mathcal{R}\, a\} = \{x \in Z \mid x - a = 3k, k \in Z\}.$$

Hence,

$$\overline{2} = \{x \in Z \mid x - 2 = 3k, k \in Z\} = \{\ldots,-4,-1,2,5,8,\ldots\},$$

$$\overline{-3} = \{x \in Z \mid x + 3 = 3k, k \in Z\} = \{\ldots,-6,-3,0,3,6,\ldots\},$$

$$\overline{5} = \{x \in Z \mid x - 5 = 3k, k \in Z\} = \{\ldots, -1,2,5,8,11,\ldots\}.$$

c. In part (d), where for (a,b), (c,d) $\in$ S = $\{(x,y) \mid x, y \in Z, y \neq 0\}$, (a,b) $\mathcal{R}$ (c,d) $\Leftrightarrow$ ad = bc, we have for each (a,b) $\in$ S,

$$\overline{(a,b)} = \{(x,y) \in S \mid (x,y) \,\mathcal{R}\, (a,b)\} = \{(x,y) \in S \mid xb = ya\}.$$

Hence, $\overline{(2,-3)} = \{(x,y) \in S \mid x(-3) = y(2)\}$ which is the set of all points (x,y), y $\neq$ 0, that lie on the line -3x = 2y and have integer coordinates. Therefore,

$$\overline{(2,-3)} = \{\ldots,(-4,6),(-2,3),(2,-3),(4,-6),(6,-9),\ldots\}.$$

The following theorem sets forth the fundamental properties of equivalent sets determined by an equivalence relation $\mathcal{R}$ defined on a set A. The results of this theorem will be used in establishing the connection between an equivalence relation and a partition of a set.

Theorem 4.4.1

Let $\mathcal{R}$ be an equivalence relation on A.

a. $\forall a \in A, \overline{a} \subseteq A$

b. $\forall a \in A, a \in \overline{a}$

c. For x, y $\in$ A, if x, y $\in \overline{a}$, then x $\mathcal{R}$ y.

d. If a, b $\in$ A, then $\overline{a} = \overline{b} \Leftrightarrow a \,\mathcal{R}\, b$.

e. If a, b $\in$ A, then either $\overline{a} = \overline{b}$ or $\overline{a} \cap \overline{b} = \emptyset$.

Proof

Let $\mathcal{R}$ be an equivalence relation on A. We shall prove parts (d) and (e), and leave the proofs of parts (a), (b) and (c) as exercises.

d. Let $a, b \in A$.

($\Rightarrow$) Suppose $\overline{a} = \overline{b}$. We must show $a \mathcal{R} b$. By part (b), $a \in \overline{a}$. Hence, $a \in \overline{b}$ since $\overline{a} = \overline{b}$. Thus, $a \mathcal{R} b$.

($\Leftarrow$) Suppose $a \mathcal{R} b$. We must show $\overline{a} = \overline{b}$; i.e., $\overline{a} \subseteq \overline{b}$ and $\overline{b} \subseteq \overline{a}$.

Let $x \in \overline{a}$. Then $x \mathcal{R} a$. Thus, we have $x \mathcal{R} a$ and $a \mathcal{R} b$. Hence, $x \mathcal{R} b$ since $\mathcal{R}$ is transitive. Therefore, $x \in \overline{b}$ and $\overline{a} \subseteq \overline{b}$.

Let $x \in \overline{b}$. Then $x \mathcal{R} b$. Now $a \mathcal{R} b \Rightarrow b \mathcal{R} a$ since $\mathcal{R}$ is symmetric. Therefore, $x \mathcal{R} b$ and $b \mathcal{R} a \Rightarrow x \mathcal{R} a$ since $\mathcal{R}$ is transitive. Therefore $x \in \overline{a}$ and $\overline{b} \subseteq \overline{a}$.

Hence $\overline{a} = \overline{b}$. ■

e.

Discussion

The proof of part (e) involves a statement of the form $p \Rightarrow (q \vee r)$. Recall from Exercises 1.2, problem (7), that $[p \Rightarrow (q \vee r)] \equiv [(p \wedge \sim q) \Rightarrow r] \vee [(p \wedge \sim r) \Rightarrow q]$ and that $[(p \wedge \sim q) \Rightarrow r] \vee [(p \wedge \sim r) \Rightarrow q]$ is true if either $(p \wedge \sim q) \Rightarrow r$ or $(p \wedge \sim r) \Rightarrow q$ is true. Hence, to show $(a,b \in A) \Rightarrow [(\overline{a} = \overline{b} \vee (\overline{a} \cap \overline{b} = \emptyset)]$ is true, we must show either $[((a,b) \in A) \wedge (\overline{a} \neq \overline{b})] \Rightarrow (\overline{a} \cap \overline{b} = \emptyset)$ is true or $[(a,b \in A) \wedge (\overline{a} \cap \overline{b} \neq \emptyset)] \Rightarrow (\overline{a} = \overline{b})$ is true. We shall show the latter.

Let $a, b \in A$ and $\overline{a} \cap \overline{b} \neq \emptyset$. We wish to show $\overline{a} = \overline{b}$. Since $\overline{a} \cap \overline{b} \neq \emptyset$, there exists $x \in \overline{a} \cap \overline{b}$. Hence, $x \in \overline{a}$ and $x \in \overline{b}$. We have $x \mathcal{R} a$ and $x \mathcal{R} b$ or $a \mathcal{R} x$ and $x \mathcal{R} b$. Thus, $a \mathcal{R} b$ and by part (d), $\overline{a} = \overline{b}$.

Therefore, either $\overline{a} \cap \overline{b} = \emptyset$ or $\overline{a} = \overline{b}$. ■

Example 4.4.2

Refer to Example 4.4.1, part (b). Note that

a. $\overline{2}, \overline{-3}, \overline{5} \subseteq Z$.

b. $2 \in \overline{2}$, $-3 \in \overline{-3}$, $5 \in \overline{5}$.

c. $-1, 8 \in \overline{2}$ and $(-1) \mathcal{R} 8$ since $(-1) - (8) = -9 = 3(-3)$.

d. $\overline{2} = \overline{5}$ since $5 \in \overline{2}$ and hence $5 \mathcal{R} 2$ while $\overline{2} \neq \overline{-3}$ since $-3 \notin \overline{2}$ and hence $(-3) \not\mathcal{R} 2$.

e. $\overline{2} = \overline{5}$ since $2 \mathcal{R} 5$ and $\overline{2} \cap \overline{-3} = \emptyset$ since $2 \not\mathcal{R} (-3)$.

Each equivalence relation $\mathcal{R}$ defined on a given set A will generate a family of subsets of A, i.e., $\mathcal{A} = \{\overline{a} \mid a \in A\}$.

Note that part (c) of Theorem 4.4.1 states that if $x, y \in \overline{a}$, not only are x and y related to a but they are related to each other. Hence, for a given equivalence relation $\mathcal{R}$ defined on A and for a fixed element $a \in A$, the elements of the equivalence set $\overline{a}$ may be considered as equivalent elements of A.

However, as previously stated, different equivalent relations defined on the same set may not generate the same family of subsets as the following example exhibits.

Example 4.4.3

Let **Z** be the set of integers and for $a, b \in \mathbf{Z}$, let:

$$a \mathcal{R}_1 b \Leftrightarrow a - b = 3k, k \in \mathbf{Z}$$
$$a \mathcal{R}_2 b \Leftrightarrow a - b = 4k, k \in \mathbf{Z}$$

Now $\mathcal{R}_1$ and $\mathcal{R}_2$ are both equivalent relations on **Z**. Let $\mathcal{A}_1$ and $\mathcal{A}_2$ be the family of subsets generated by $\mathcal{R}_1$ and $\mathcal{R}_2$, respectively. For $2 \in \mathbf{Z}$, we have the equivalent sets:

$$\begin{aligned} \overline{2} &= \{x \in \mathbf{Z} \mid x \mathcal{R}_1 2\} \\ &= \{x \in \mathbf{Z} \mid x - 2 = 3k, k \in \mathbf{Z}\} \\ &= \{\ldots, -4, -1, 2, 5, 8, \ldots\} \in \mathcal{A}_1. \end{aligned}$$

Example 4.4.3 continued next page.

Example 4.4.3 Continued

$$\begin{aligned}\overline{2} &= \{x \in \mathbf{Z} \mid x\, \mathcal{R}_2 2\} \\ &= \{x \in \mathbf{Z} \mid x - 2 = 4k,\ k \in \mathbf{Z}\} \\ &= \{\,\ldots, -6, -2, 2, 6, 10, \ldots\,\} \in \mathcal{A}_2.\end{aligned}$$

Hence, we see that the equivalence set $\overline{2}$ determined by $\mathcal{R}_1$ and the equivalence set $\overline{2}$ determined by $\mathcal{R}_2$ are not the same and thus $\mathcal{A}_1 \neq \mathcal{A}_2$. Furthermore, note that under the relation $\mathcal{R}_1$, the integers 2 and 5 may be considered equivalent since $5 \in \overline{2}$; but under the relation $\mathcal{R}_2$, 2 and 5 may not be considered equivalent elements since $5 \notin \overline{2}$.

Exercises 4.4

1. Prove parts (a), (b), and (c) of Theorem 4.4.1.

2. Let $A = \{1,2,3,4\}$ and $\mathcal{R} = \{(1,1),(2,2),(3,3),(4,4),(1,3),(2,4),(3,1),(4,2)\}$.

   a. Verify that $\mathcal{R}$ is an equivalence relation on A.

   b. For each element in A, find the corresponding equivalence set determined by $\mathcal{R}$.

Section 4.5
PARTITIONS

We are all familiar with the partitions of a house that create separate rooms. These rooms are disjoint, each room has some area of space and, when taken collectively, they form the entire house. This concept can be extended to the partition of a set. As we shall see, the concepts of a partition and an equivalence relation are closely connected and, in essence, are slightly different ways of obtaining similar results.

Definition 4.5.1

Let $A \neq \emptyset$ and $\{X_i \mid i \in \mathbf{I}\}$ be a non-empty indexed family of subsets of A. Then $\{X_i \mid i \in \mathbf{I}\}$ is called a ***partition*** of A if:

a. $X_i \neq \emptyset, \forall i \in \mathbf{I}$.

b. $A = \bigcup_{i \in \mathbf{I}} X_i$

c. $X_i, X_j \in \{X_i \mid i \in \mathbf{I}\}$, then either $X_i = X_j$ or $X_i \cap X_j = \emptyset$; i.e., $\{X_i \mid i \in \mathbf{I}\}$ is pairwise disjoint.

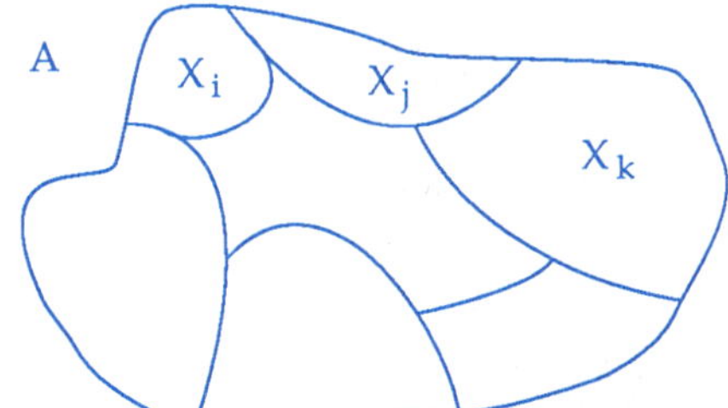

Example 4.5.1

a. Let $A = \{a,b,c,d,e\}$ and $X_1 = \{a,c\}, X_2 = \{b,e\}, X_3 = \{d\}$. Then the non-empty family of subsets, $\{X_1,X_2,X_3\}$, of A is a partition of A since

   i. $X_i \neq \emptyset, \forall i \in \{1,2,3\}$
   ii. $A = X_1 \cup X_2 \cup X_3$
   iii. $\{X_1,X_2,X_3\}$ is pairwise disjoint.

b. For each $i \in \mathbf{Z}$, let $X_i = [i, i + 1) = \{x \in \mathbf{R} \mid i \leq x < i + 1\}$. Then the non-empty family of subsets, $\{X_i \mid i \in \mathbf{Z}\}$, of $\mathbf{R}$ is a partition of $\mathbf{R}$.

   i. $X_i \neq \emptyset, \forall i \in \mathbf{Z}$ since $i \in X_i$.

   ii. $\mathbf{R} = \bigcup_{i \in \mathbf{Z}} X_i$.

   Since $\forall i \in \mathbf{Z}, X_i \subseteq \mathbf{R}$, $\bigcup_{i \in \mathbf{Z}} X_i \subseteq \mathbf{R}$ by Exercises 2.4, problem (3). We must thus show $\mathbf{R} \subseteq \bigcup_{i \in \mathbf{Z}} X_i$. Let $x \in \mathbf{R}$. If x is an integer, let $x = i$. Then $x \in [i, i + 1) = X_i$. If x is not an integer, then $\exists i \in \mathbf{Z}$ such that $i < x < i + 1$ and $x \in [i, i + 1) = X_i$. In each case, we have $x \in X_i \subseteq \bigcup_{i \in \mathbf{Z}} X_i$. Therefore, $\mathbf{R} \subseteq \bigcup_{i \in \mathbf{Z}} X_i$. Hence, $\mathbf{R} = \bigcup_{i \in \mathbf{Z}} X_i$.

Example 4.5.1 continued next page.

Example 4.5.1 Continued

iii. For all i, j $\in$ **Z**, if $i \neq j$, then $X_i \cap X_j = \emptyset$; i.e., $\{X_i \mid i \in \mathbf{Z}\}$ is pairwise disjoint.

c. $A = \{1,2,3,4,5,6\}$.

i. If $X_1 = \{1,3,6\}$, $X_2 = \{2,4\}$, $X_3 = \{1,5\}$, then $\{X_1,X_2,X_3\}$ is not a partition of A since $X_1 \cap X_3 \neq \emptyset$ but $X_1 \neq X_3$; i.e., $\{X_1,X_2,X_3\}$ is not pairwise disjoint.

ii. If $Y_1 = \{1,3\}$, $Y_2 = \{2\}$ and $Y_3 = \{4,6\}$, then $\{Y_1,Y_2,Y_3\}$ is not a partition of A since $A \neq \bigcup_{i=1}^{3} Y_i$.

Now let us return to an equivalence relation and the equivalence sets determined by that relation. By referring to Theorem 4.4.1, we see that these equivalence sets exhibit many of the properties required for a family of subsets to form a partition of a given set. This connection between equivalence sets and a partition is formalized in the following theorem.

Theorem 4.5.1

Let $\mathcal{R}$ be an equivalence relation on A and $\{\overline{a} \mid a \in A\}$ the collection of all equivalence sets of A determined by $\mathcal{R}$. Then $\{\overline{a} \mid a \in A\}$ is a partition of A.

Proof

Let $\mathcal{R}$ be an equivalence relation on A and $\{\overline{a} \mid a \in A\}$ be the collection of all equivalence sets determined by $\mathcal{R}$.

a. Since $a \in \overline{a}$, $\forall a \in A$, we have $\overline{a} \neq \emptyset$, $\forall a \in A$.

b. $A = \bigcup_{a \in A} \overline{a}$.

Let $x \in A \Rightarrow x \in \overline{x}$.

Now $\overline{x} \in \{\overline{a} \mid a \in A\} \Rightarrow \overline{x} \subseteq \bigcup_{a \in A} \overline{a} \Rightarrow x \in \bigcup_{a \in A} \overline{a}$.

Therefore, $A \subseteq \bigcup_{a \in A} \bar{a}$. Let $x \in \bigcup_{a \in A} \bar{a} \Rightarrow x \in \bar{a}$ for some $a \in A$.

Now $\bar{a} \subseteq A \Rightarrow x \in A$. Therefore, $\bigcup_{a \in A} \bar{a} \subseteq A$. Hence, $A = \bigcup_{a \in A} \bar{a}$.

c. $\{\bar{a} \mid a \in A\}$ is pairwise disjoint.

Let $\bar{x}, \bar{y} \in \{\bar{a} \mid a \in A\}$. By Theorem 4.4.1, part (e), $\bar{x} = \bar{y}$ or $\bar{x} \cap \bar{y} = \emptyset$, i.e., $\{\bar{a} \mid a \in A\}$ is pairwise disjoint.

Hence, $\{\bar{a} \mid a \in A\}$ is a partition of A. ■

In the context of a fixed equivalence relation $\mathcal{R}$ on A, the partition $\{\bar{a} \mid a \in A\}$ is referred to as the partition induced or defined by the equivalence relation $\mathcal{R}$.

Example 4.5.2

a. Let $A = \{a,b,c,d\}$ and $\mathcal{R} = \{(a,a),(b,b),(c,c),(d,d),(b,d),(d,b)\}$. Then $\mathcal{R}$ is an equivalence relation on A; and, as shown in Example 4.4.1, part (a), $\bar{a} = \{a\}$, $\bar{b} = \{b,d\}$, $\bar{c} = \{c\}$, $\bar{d} = \{d,b\}$. Hence $\{\bar{a},\bar{b},\bar{c},\bar{d}\} = \{\bar{a},\bar{b},\bar{c}\}$ is a partition of A.

b. Refer to Example 4.4.1, part (b), where for a, b, $\in$ **Z**, $a \mathcal{R} b \Leftrightarrow a - b = 3k$, $k \in$ **Z**. Now the collection of all equivalence sets determined by $\mathcal{R}$; i.e., $\{\ldots,\overline{-3},\overline{-2},\overline{-1},\bar{0},\bar{1},\bar{2},\bar{3},\ldots\}$, forms a partition of **Z**. We shall show in Chapter 6 that there are only three distinct equivalence sets of **Z** determined by $\mathcal{R}$ and, as a consequence, the partition of **Z** can be represented by the family of sets $\{\bar{0},\bar{1},\bar{2}\}$.

Conversely, a partition of a set A will induce an equivalence relation on A. This result is proven in the following theorem.

Theorem 4.5.2

Let $\{X_i \mid i \in \mathbf{I}\}$ be a non-empty family of subsets of A. If $\{X_1 \mid i \in \mathbf{I}\}$ is a partition of A, then $\{X_i \mid i \in \mathbf{I}\}$ induces an equivalence relation on A.

Proof

Let $\{X_i \mid i \in \mathbf{I}\}$ be a partition of A. For a, b ∈ A, let $a \mathcal{R} b \Leftrightarrow \exists X_i \in \{X_i \mid i \in \mathbf{I}\}$ such that $a, b \in X_i$. In other words, a and b will be related if and only if they belong to the same element, X_i, of the partition. We shall show $\mathcal{R}$ is an equivalence relation on A.

Let $a \in A$. Since $A = \bigcup_{i \in \mathbf{I}} X_i$, $a \in \bigcup_{i \in \mathbf{I}} X_i \Rightarrow a \in X_i$ for some $i \in \mathbf{I}$. Hence $a \mathcal{R} a$ and thus $\mathcal{R}$ is reflexive.

Let $a \mathcal{R} b$. Then $\exists X_i \in \{X_i \mid i \in \mathbf{I}\}$ such that $a, b \in X_i$. Hence $b, a \in X_i \Rightarrow b \mathcal{R} a$ and thus $\mathcal{R}$ is symmetric.

Let $a \mathcal{R} b$ and $b \mathcal{R} c$. Then $\exists X_i, X_j \in \{X_i \mid i \in \mathbf{I}\}$ such that $a, b \in X_i$ and $b, c \in X_j$. Since $b \in X_i \cap X_j$, $X_i \cap X_j \neq \emptyset$. Therefore $X_i = X_j$. Hence $a, c \in X_i \Rightarrow a \mathcal{R} c$ and thus $\mathcal{R}$ is transitive.

Since $\mathcal{R}$ is reflexive, symmetric and transitive, $\mathcal{R}$ is an equivalence relation on A. ■

Example 4.5.3

Let $A = \{a,b,c,d,e\}$ and $X_1 = \{a,c\}$, $X_2 = \{b,e\}$, $X_3 = \{d\}$. Then $\{X_1,X_2,X_3\}$ is a partition of A. For $x, y \in A$, $x \mathcal{R} y \Leftrightarrow \exists X_i \in \{X_1,X_2,X_3\}$ such that $x, y \in X_i$. By Theorem 4.5.2, $\mathcal{R}$ is an equivalence relation on A.

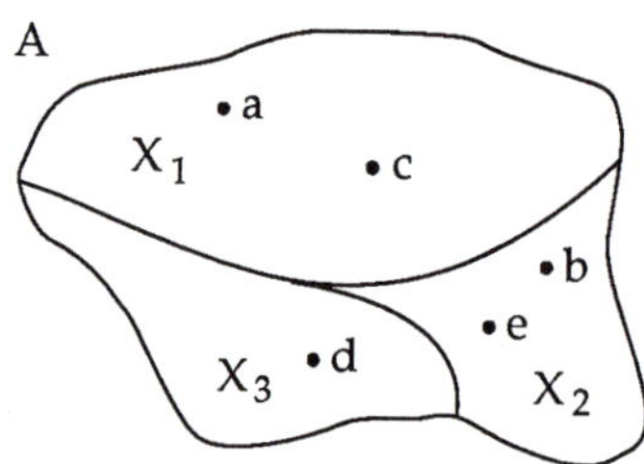

a. Listing the elements of the relation $\mathcal{R}$, we have,

$\mathcal{R} = \{(a,a),(c,c),(a,c),(c,a),(b,b),(e,e),(b,e),(e,b),(d,d)\}$.

b. For each $x \in A$, the equivalence set $\bar{x}$ determined by $\mathcal{R}$ is:

$\bar{a} = \{x \in A \mid x \mathcal{R} a\} = \{x \in A \mid (x,a) \in \mathcal{R}\} = \{a,c\} = X_1$

$\bar{b} = \{x \in A \mid x \mathcal{R} b\} = \{x \in A \mid (x,b) \in \mathcal{R}\} = \{b,e\} = X_2$

$\bar{c} = \{x \in A \mid x \mathcal{R} c\} = \{x \in A \mid (x,c) \in \mathcal{R}\} = \{c,a\} = X_1$

$\bar{d} = \{x \in A \mid x \mathcal{R} d\} = \{x \in A \mid (x,d) \in \mathcal{R}\} = \{d\} = X_3$

$\bar{e} = \{x \in A \mid x \mathcal{R} e\} = \{x \in A \mid (x,e) \in \mathcal{R}\} = \{e,b\} = X_2$

Thus the equivalence sets of A determined by $\mathcal{R}$ are the sets of the given partition of A.

Let A be an non-empty set. We have shown that each partition on A induces an equivalence relation on A and that each equivalence relation on A induces a partition on A. We will leave as an exercise (Exercises 4.5, problem (4)) the proof that there is a 1–1 correspondence between the set of all equivalence relations on A and the set of all partitions on A.

Exercises 4.5

1. By Theorem 4.5.1, the collection of all equivalence sets forms a partition of A. For Exercises 4.4, problem (2), find the partition of A determined by $\mathcal{R}$.

2. Let $A = \{a,b,c,d,e\}$.

   a. Find three (3) different partitions of A.

   b. By Theorem 4.5.2, each partition will induce an equivalence relation $\mathcal{R}$ on A. For each partition found in part (a), list the elements of the corresponding induced relation $\mathcal{R}$.

3. Let $A = \{1,2,3,4,5,6\}$ and $\mathcal{R} = \{(1,1),(1,5),(2,2),(2,6),(3,3),(3,4),(4,4),(4,3),(5,5),(5,1),(6,6),(6,2)\}$.
   a. Show $\mathcal{R}$ is an equivalence relation on A.
   b. Find the partition of A induced by $\mathcal{R}$.

4. Let θ denote the set of all equivalence relations on a non-empty set A and ψ denote the set of all partitions of A. Define $f: \theta \to \psi$ which associates each $\mathcal{R} \in \theta$ with its induced partition P in ψ.
   a. Show that f is a function from θ to ψ.
   b. Show that f is onto.
   c. Show that f is 1–1.

Chapter 5

BINARY OPERATIONS

We are familiar with the terms "addition," "multiplication," and "subtraction" as used in the set of integers, Z, as well as the terms "union," "intersection," and "difference" as used in the power set P(X). These are but a few examples of a more general concept referred to as a *binary operation* on a set. We shall not only look at the concept of binary operations but also some important properties which they may possess.

Section 5.1
BINARY OPERATIONS

Intuitively, a binary operation on a set A is some type of a rule whereby any two elements in A are combined to produce another element in A. We now state the formal definition of a binary operation.

Definition 5.1.1

Let $A \neq \emptyset$. A ***binary operation*** (or simply an operation) on the set A is a function from A x A to A.

Symbols such as $*$, #, o, +, •, etc., will be used to denote operations on a set. Let $*$ be an operation on A. Then by definition, $*$ is a function from A x A to A; i.e., $*$ is a correspondence that assigns to each element $(a,b) \in A \times A$ an unique element $*(a,b) \in A$. When $*$ is an operation, it is customary to write the image of (a,b) under $*$ as $a * b$ rather than $*(a,b)$. Thus, $*: A \times A \to A$; and, if $(a,b) \in A \times A$, then $a * b \in A$.

Example 5.1.1

Let $A = \{1,2,3\}$.
Then $A \times A = \{(1,1),(1,2),(1,3),(2,1),(2,2),(2,3),(3,1),(3,2),(3,3)\}$

Now any function from A x A to A will yield an operation on A.

a. Consider the function $*: A \times A \to A$ where

(1,1) → 2	$1 * 1 = 2$
(1,2) → 2	$1 * 2 = 2$
(1,3) → 2	$1 * 3 = 2$
(2,1) → 3	$2 * 1 = 3$
(2,2) → 3	$2 * 2 = 3$
(2,3) → 3	$2 * 3 = 3$
(3,1) → 1	$3 * 1 = 1$
(3,2) → 1	$3 * 2 = 1$
(3,3) → 1	$3 * 3 = 1$

Now $*$ is an operation on A which can conveniently be represented in table form as:

$*$	1	2	3
1	2	2	2
2	3	3	3
3	1	1	1

b. Consider the function #: A x A → A where

(1,1) → 1		1 # 1 = 1
(1,2) → 2		1 # 2 = 2
(1,3) → 3		1 # 3 = 3
(2,1) → 2		2 # 1 = 2
(2,2) → 2		2 # 2 = 2
(2,3) → 2		2 # 3 = 2
(3,1) → 1		3 # 1 = 1
(3,2) → 3		3 # 2 = 3
(3,3) → 3		3 # 3 = 3

Then # is an operation on A which can conveniently be represented in table form as:

#	1	2	3
1	1	2	3
2	2	2	2
3	1	3	3

c. How many operations can be defined on A? Since A x A has 9 elements and A has 3 elements, the number of functions from A x A → A is 3^9. Hence, there are 3^9 = 19,683 distinct operations that can be defined on A.

Example 5.1.2

a. Let Z be the set of integers. Then $+: (a,b) \rightarrow a + b$ and $\bullet: (a,b) \rightarrow a \bullet b$ are operations on Z.

b. Let X be a set and P(X) the power set of X. Then $\cup: (A,B) \rightarrow A \cup B$; $\cap: (A,B) \rightarrow A \cap B$ and $-: (A,B) \rightarrow A - B$ are operations on P(X).

If $X = \{a,b\}$, then $P(X) = \{\emptyset,\{a\},\{b\},X\}$. For the operations of $\cup$ and $-$ we have the following tables: (Note: A review of Theorems 2.3.3 and 2.3.5 in Chapter 2 may be helpful.)

$\cup$	$\emptyset$	{a}	{b}	X
$\emptyset$	$\emptyset$	{a}	{b}	X
{a}	{a}	{a}	X	X
{b}	{b}	X	{b}	X
X	X	X	X	X

$-$	$\emptyset$	{a}	{b}	X
$\emptyset$	$\emptyset$	$\emptyset$	$\emptyset$	$\emptyset$
{a}	{a}	$\emptyset$	{a}	$\emptyset$
{b}	{b}	{b}	$\emptyset$	$\emptyset$
X	X	{b}	{a}	$\emptyset$

c. Let $A \neq \emptyset$ and $F = \{f \mid f: A \rightarrow A\}$. Then the composition $\circ: (f,g) \rightarrow f \circ g$, is an operation on F.

If $A = \{a,b\}$, then $F = \{\mathbf{e},f,g,h\}$ where

e: $a \rightarrow a$, $b \rightarrow b$

f: $a \rightarrow b$, $b \rightarrow a$

g: $a \rightarrow a$, $b \rightarrow a$

h: $a \rightarrow b$, $b \rightarrow b$

For the operation $\circ$ we have the following table:

o	**e**	f	g	h
e	**e**	f	g	h
f	f	**e**	h	g
g	g	g	g	g
h	h	h	h	h

d. Let $\mathbf{R}^* = \mathbf{R} - \{0\}$; i.e., $\mathbf{R}^*$ is the set of non-zero real numbers. For $a, b \in \mathbf{R}^*$, let $a \# b = |a| b$. Since $a, b \in \mathbf{R}^*$, we have $|a|, b \in \mathbf{R}^*$ and hence $|a| b \in \mathbf{R}$. Thus $a \# b \in \mathbf{R}^*$ and $\#: \mathbf{R}^* \times \mathbf{R}^* \rightarrow \mathbf{R}^*$; i.e., # is an operation on $\mathbf{R}^*$.

e. For m, n $\in$ **N**, let $m * n = 2^m \cdot (2n - 1)$. Since m, n $\in$ **N**, we have 2^m, $2n - 1 \in$ **N** and hence $2^m \cdot (2n - 1) \in$ **N**. Thus $m * n \in$ **N** and $*: \mathbf{N} \times \mathbf{N} \to \mathbf{N}$; i.e., $*$ is an operation on **N**.

f. For x, y $\in$ **R**, let $x * y = \sqrt{x + y}$. Now $*$ is not an operation on **R** since $*$ is not a function from **R** x **R** $\to$ **R**. To see this, consider x = -4 and y = 3. Then $(-4,3) \in \mathbf{R} \times \mathbf{R}$, but $(-4) * 3 = \sqrt{-4 + 3} = \sqrt{-1} \notin \mathbf{R}$. Thus, if $(x,y) \in \mathbf{R} \times \mathbf{R}$, $x * y$ may not be in **R**.

The definition of operation implicitly implies two properties: *closure* and *well-defined.* Let us assume that $*$ is an operation on A.

Closure Property: If a, b $\in$ A, then $a * b \in$ A.

Well-defined Property: Since $*$ is a function from A x A to A, each element of A x A must be assigned to one and only one element of A. Thus, if some elements of A can be represented in more than one way, it must be verified that the same result is obtained no matter which representation is used. That is, if $a = a'$ and $b = b'$ (and hence $(a, b) = (a', b')$), it must be shown that $a * b = a' * b'$.

Example 5.1.3

Consider the set of rational numbers, $\mathbf{Q} = \{\frac{a}{b} \mid a, b \in \mathbf{Z} \text{ and } b \neq 0\}$.

For $\frac{a}{b}, \frac{c}{d} \in \mathbf{Q}$, $\frac{a}{b} \cdot \frac{c}{d} = \frac{ac}{bd}$. Since ac, bd $\in$ **Z** and bd $\neq 0$, $\frac{ac}{bd} \in \mathbf{Q}$.

Thus **Q** is closed relative to multiplication. Now each rational number has more than one representation. For example $\frac{2}{5} = \frac{6}{15}$ and $\frac{-4}{3} = \frac{8}{-6}$. Hence, if $\frac{a}{b} = \frac{p}{q}$ and $\frac{c}{d} = \frac{r}{s}$, it must be verified that $\frac{a}{b} \cdot \frac{c}{d} = \frac{p}{q} \cdot \frac{r}{s}$ or $\frac{ac}{bd} = \frac{pr}{qs}$. If we can show that $(ac)(qs) = (bd)(pr)$, then we shall have verified that multiplication is well-defined.

Now

$$\frac{a}{b} = \frac{p}{q} \Rightarrow aq = bp \text{ and } \frac{c}{d} = \frac{r}{s} \Rightarrow cs = dr$$

Hence,
$$\begin{aligned}(ac)(qs) &= (aq)(cs)\\ &= (bp)(dr)\\ &= (bd)(pr)\end{aligned}$$

Exercises 5.1

For each of the following, determine whether or not # is an operation on the given set.

1. $x \# y = \frac{x+1}{y^2+1}$ where $x, y \in \mathbf{R}$.
2. $x \# y = \frac{x+1}{y^2-1}$ where $x, y \in \mathbf{R}$.
3. $x \# y = x\,e^y$ where $x, y \in \mathbf{R}^+ = \{a \in \mathbf{R} \mid a > 0\}$.
4. $x \# y = \frac{x}{y}$ where $x, y \in \mathbf{Z}^* = \mathbf{Z} - \{0\}$.
5. $x \# y = 2^{x+y}$ where $x, y \in \mathbf{Z}$.

Section 5.2
PROPERTIES OF OPERATIONS

Operations that possess special properties are the ones that are of most importance in mathematics. In this section we introduce some of these special properties.

Definition 5.2.1

Let $*$ be an operation on A.

a. The operation $*$ is ***commutative*** if $\forall a, b \in A$, $a * b = b * a$.
b. The operation $*$ is ***associative*** if $\forall a, b, c \in A$, $a * (b * c) = (a * b) * c$.

Let $*$ be an operation on the set A. To determine whether or not $*$ is commutative, select two arbitrary elements $a, b \in A$ and compute $a * b$ and $b * a$. If $a * b = b * a$, then $*$ is commutative. If $a * b \neq b * a$, then $*$ is not commutative. To determine whether or not $*$ is associative, select three arbitrary elements $a, b, c \in A$ and compute $a * (b * c)$ and $(a * b) * c$. If $a * (b * c) = (a * b) * c$, then $*$ is associative; and, if $a * (b * c) \neq (a * b) * c$, then $*$ is not associative.

Example 5.2.1

Refer to Example 5.1.2.

a. The operations + and • are both commutative and associative operations on **Z**. That is $\forall a, b \in \mathbf{Z}$, $a + b = b + a$ and $ab = ba$; and, $\forall a, b, c \in \mathbf{Z}$, $a + (b + c) = (a + b) + c$ and $a(bc) = (ab)c$.

b. The operations $\cup$ and $\cap$ are both commutative and associative operations on P(X). The operation – need not be a commutative nor an associative operation on P(X). (See Table, Example 5.1.2, part (b).)

   Let $X = \{a,b\}$. Then $P(X) = \{\emptyset, \{a\}, \{b\}, X\}$.

   Now $\{a\} - \{b\} = \{a\}$ and $\{b\} - \{a\} = \{b\}$. Hence, $\{a\} - \{b\} \neq \{b\} - \{a\}$.

   Also, $\{a\} - (\{b\} - \{a\}) = \{a\} - \{b\} = \{a\}$ and $(\{a\} - \{b\}) - \{a\} = \{a\} - \{a\} = \emptyset$. Hence, $\{a\} - (\{b\} - \{a\}) \neq (\{a\} - \{b\}) - \{a\}$.

c. The operation $\circ$ on $F = \{f \mid f: A \to A\}$ is an associative operation since by Theorem 3.3.1, $f \circ (g \circ h) = (f \circ g) \circ h \ \forall f, g, h \in F$. However $\circ$ need not be a commutative operation. (See Table, Example 5.1.2, part (c).)

   Let $A = \{a,b\}$, $g: A \to A$ where $g(a) = a$, $g(b) = a$ and $h: A \to A$ where $h(a) = b$, $h(b) = b$. Then $g \circ h \neq h \circ g$.

d. The operation # on $\mathbf{R}^* = \mathbf{R} - \{0\}$ where $a \# b = |a|b$ is not a commutative operation since $(-2) \# 3 \neq 3 \# (-2)$. However, the operation is associative. Let $a, b, c \in \mathbf{R}^*$.

$$\begin{aligned}(a \# b) \# c &= |a \# b|\, c \\ &= \big|\,|a|b\,\big|\, c \\ &= (\,||a||\,|b|\,)\, c \\ &= (\,|a|\,|b|\,)\, c\end{aligned} \qquad \begin{aligned}a \# (b \# c) &= |a|\,(b \# c) \\ &= |a|\,(\,|b|\, c) \\ &= (\,|a|\,|b|)\, c\end{aligned}$$

   Therefore, $(a \# b) \# c = a \# (b \# c)$.

e. Let $A = \{x \mid x = 2^m, m \in \mathbf{Z}\} = \{\ldots, 2^{-3}, 2^{-2}, 2^{-1}, 2^0, 2^1, 2^2, 2^3, \ldots\}$ For $x = 2^m$, $y = 2^n \in A$, let $x * y = 2^m * 2^n = 2^{m+n}$. Then $*$ is an operation on A. (Keep in mind + is a commutative and an associative operation on **Z**.)

Example 5.2.1 continued next page.

Example 5.2.1 Continued

Let $x, y \in A$. Then $x = 2^m$ and $y = 2^n$ for $m, n \in \mathbf{Z}$.

$$x * y = 2^m * 2^n = 2^{m+n}$$

$$y * x = 2^n * 2^m = 2^{n+m} = 2^{m+n}$$

Since $x * y = y * x$, $*$ is commutative.

Let $x, y, z \in A$. Then $x = 2^m$, $y = 2^n$ and $z = 2^p$ for $m, n, p \in \mathbf{Z}$.

$$x * (y * z) = 2^m * (2^n * 2^p) = 2^m * 2^{n+p} = 2^{m+(n+p)}$$

$$(x * y) * z = (2^m * 2^n) * 2^p = 2^{m+n} * 2^p = 2^{(m+n)+p} = 2^{m+(n+p)}$$

Since $x * (y * z) = (x * y) * z$, $*$ is associative.

Definition 5.2.2

Let # and $*$ be two operations on the set A. The operation # is ***distributive*** over the operation $*$ if $\forall a, b, c \in A$, $a \# (b * c) = (a \# b) * (a \# c)$ and $(b * c) \# a = (b \# a) * (c \# a)$.

Theorem 5.2.1

Let # and $*$ be two operations on the set A.
If $\forall a, b, c \in A$, $a \# (b * c) = (a \# b) * (a \# c)$ and
if # is commutative, then # is distributive over $*$.

Proof

Let # be commutative and $\forall a, b, c \in A$, $a \# (b * c) = (a \# b) * (a \# c)$.

Then $(b * c) \# a = a \# (b * c)$
$= (a \# b) * (a \# c)$
$= (b \# a) * (c \# a)$.

Hence, # is distributive over $*$. ■

Example 5.2.2

Refer to Example 5.1.2.

a. In **Z**:

i. $\bullet$ is distributive over + since $\forall a, b, c \in \mathbf{Z}$, $a(b + c) = ab + ac$ and $\bullet$ is commutative.

ii. + is not distributive over $\bullet$ since $2 + [(3)(4)] \neq (2 + 3)(2 + 4)$.

b. In P(X):

i. $\cup$ is distributive over $\cap$ and $\cap$ is distributive over $\cup$ since $\forall A, B, C \in P(X)$,

$$A \cup (B \cap C) = (A \cup B) \cap (A \cup C),$$
$$A \cap (B \cup C) = (A \cap B) \cup (A \cap C)$$ and

both $\cup$ and $\cap$ are commutative (see Theorem 2.3.4).

ii. $\cap$ is distributive over $-$ since $\forall A, B, C \in P(X)$,

$$A \cap (B - C) = (A \cap B) - (A \cap C)$$

and $\cap$ is commutative (see Theorem 2.3.7).

iii. $\cup$ need not be distributive over $-$ (see Tables, Example 5.1.2, part (b)).

Let $X = \{a,b\}$. Then $P(X) = \{\emptyset,\{a\},\{b\},X\}$.
Now $\{a\} \cup (\{a\} - \{b\}) = \{a\} \cup \{a\} = \{a\}$
while $(\{a\} \cup \{a\}) - (\{a\} \cup \{b\}) = \{a\} - \{a,b\} = \{a\} - X = \emptyset$.
Thus, $\{a\} \cup (\{a\} - \{b\}) \neq (\{a\} \cup \{a\}) - (\{a\} \cup \{b\})$.

Exercises 5.2

For 1-5 determine whether or not the operation(s) defined on the given set is (are) commutative and/or associative.

1. For $x, y \in \mathbf{R} - \{1\}$, let $x * y = x + y - xy$.

2. For $x, y \in \mathbf{R}$, let $x * y = 1 - xy$.

3. For $x = (a, b), y = (c, d) \in \mathbf{R} \times \mathbf{R}$, let
   a. $x + y = (a,b) + (c,d) = (a + c, b + d)$.
   b. $xy = (a,b)(c,d) = (ac,bd)$.

4. For $g, h \in F = \{f \mid f: \mathbf{R} \to \mathbf{R}\}$, consider
   a. $g + h$ where $(g + h)(x) = g(x) + h(x)\ \forall x \in \mathbf{R}$.
   b. gh where $(gh)(x) = g(x)h(x)\ \forall x \in \mathbf{R}$.

5. Let M be the set of all 2 x 2 matrices over **R** and $x, y \in M$ where

$$x = \begin{bmatrix} a_1 & a_2 \\ a_3 & a_4 \end{bmatrix} \quad y = \begin{bmatrix} b_1 & b_2 \\ b_3 & b_4 \end{bmatrix}$$

   a. Let $x + y = \begin{bmatrix} a_1 & a_2 \\ a_3 & a_4 \end{bmatrix} + \begin{bmatrix} b_1 & b_2 \\ b_3 & b_4 \end{bmatrix} = \begin{bmatrix} a_1 + b_1 & a_2 + b_2 \\ a_3 + b_3 & a_4 + b_4 \end{bmatrix}$

   b. Let $xy = \begin{bmatrix} a_1 & a_2 \\ a_3 & a_4 \end{bmatrix} \bullet \begin{bmatrix} b_1 & b_2 \\ b_3 & b_4 \end{bmatrix} = \begin{bmatrix} a_1b_1 + a_2b_3 & a_1b_2 + a_2b_4 \\ a_3b_1 + a_4b_3 & a_3b_2 + a_4b_4 \end{bmatrix}$

6. For 3-5 show that the operation • is distributive over the operation + (when possible, make use of Theorem 5.2.1).

Section 5.3
IDENTITY ELEMENTS

In the integers, **Z**, the numbers 0 and 1 are special elements called *identity elements*. The reason for this is that they play an important role: for any $n \in \mathbf{Z}$, $n + 0 = 0 + n = n$ and $n \cdot 1 = 1 \cdot n = n$. That is, when 0 is combined with any integer n under the operation of addition, the integer n is reproduced; and, when 1 is combined with any integer n under the operation of multiplication, the integer n is reproduced. In this section we investigate various sets with a given operation in order to determine whether or not the sets possess a special element that exhibits a property similar to the integers 0 and 1.

Definition 5.3.1

Let $*$ be an operation on the set A and $e \in A$. The element e is called an ***identity*** element of A relative to the operation $*$ if $\forall a \in A$, $a * e = a$ and $e * a = a$.

Example 5.3.1

Refer to Example 5.1.2.

a. In the set of integers **Z**:

   i. 0 is an identity element relative to the operation + since $\forall a \in \mathbf{Z}$, $a + 0 = a = 0 + a$.

   ii. 1 is an identity element relative to the operation • since $\forall a \in \mathbf{Z}$, $a \cdot 1 = a = 1 \cdot a$.

b. In P(X):

   i. $\emptyset$ is an identity element relative to the operation $\cup$ since $\forall A \in P(X)$, $A \cup \emptyset = A = \emptyset \cup A$.

   ii. X is an identity element relative to the operation $\cap$ since $\forall A \in P(X)$, $A \cap X = A = X \cap A$.

Example 5.3.1 continued next page.

Example 5.3.1 Continued

iii. P(X) need not have an identity element relative to the operation –.

Let X = {a,b}. Then P(X) = {∅,{a},{b},X}.

–	∅	{a}	{b}	X
∅	∅	∅	∅	∅
{a}	{a}	∅	{a}	∅
{b}	{b}	{b}	∅	∅
X	X	{b}	{a}	∅

If P(X) has an identity element e relative to – , then $\forall A \in P(X)$, $A - e = A$ and $e - A = A$. Now the equation $A - e = A$ has a solution in P(X) if $e = \emptyset$; that is, $A - \emptyset = A$, $\forall A \in P(X)$. However, $e = \emptyset$ does not satisfy the equation $e - A = A$, $\forall A \in P(X)$, since $\emptyset - A = \emptyset$. Hence, P(X) has no identity element relative to –.

c. If $F = \{f \mid f{:}A \to A\}$, the identity function **e**, where $\mathbf{e}(a) = a$, $\forall a \in A$, is an identity element relative to the operation ∘ since $\forall f \in F$, $f \circ \mathbf{e} = f = \mathbf{e} \circ f$.

i. $D_{f \circ \mathbf{e}} = D_{\mathbf{e}} = A = D_f$ $\qquad$ $D_{\mathbf{e} \circ f} = D_f$

ii. $(f \circ \mathbf{e})(a) = f(\mathbf{e}(a)) = f(a)$ $\qquad$ $(\mathbf{e} \circ f)(a) = \mathbf{e}(f(a)) = f(a)$

Therefore, $f \circ \mathbf{e} = f$. $\qquad$ Therefore, $\mathbf{e} \circ f = f$.

Hence, $f \circ \mathbf{e} = f = \mathbf{e} \circ f$.

d. $\mathbf{R}^* = \mathbf{R} - \{0\}$ does not have an identity element relative to the operation # where $a \# b = |a|b$.

If $\mathbf{R}^*$ has an identity element, e, relative to the operation #, then $\forall a \in \mathbf{R}^*$, $a \# e = a$ and $e \# a = a$. Now if $e \# a = a$, then $|e|a = a$; and, since $a \neq 0$, we have $|e| = 1$.

Therefore, the only choices for e are 1 or -1.

If $e = 1$, $1 \# a = |1|a = (1)a = a$ but $a \# 1 = |a|(1) = a$ only if $a > 0$.

If $e = -1$, $(-1) \# a = |-1| a = (1) a = a$ but $a \# (-1) = |a| (-1) = a$ only if $a < 0$.

Since neither $e = 1$ nor $e = -1$ satisfies the equation $a \# e = a \ \forall a \in \mathbf{R}^*$, $\mathbf{R}^*$ has no identity element relative to the operation #.

e. $A = \{x \mid x = 2^m, m \in Z\}$ has an identity element relative to the operation $*$ where $x * y = 2^m * 2^n = 2^{m+n}$.

If A has an identity element, e, relative to the operation $*$, then $\forall x \in A$, $x * e = x$ and $e * x = x$. Now if $x * e = x$ and if $x = 2^m$ and $e = 2^n$, then we have

$$\begin{aligned} x * e &= x \\ 2^m * 2^n &= 2^m \\ 2^{m+n} &= 2^m \\ m + n &= m \\ n &= m - m \\ n &= 0 \end{aligned}$$

Therefore, the equation $x * e = x$ has a solution in A; i.e., $e = 2^0$. Since $*$ is commutative, $e * x = x$. Hence, $e = 2^0$ is an identity element relative to $*$.

Theorem 5.3.1

Let $*$ be an operation on the set A. If A has an identity element, e, relative to the operation $*$, then e is unique.

Discussion

The proof of this theorem involves the concept of uniqueness. In such cases one wishes to show that, if an element $x \in A$ possesses a certain property, then x is the only element of A having that property. Thus, to show uniqueness, one assumes there exists two elements $x, y \in A$ having the given property and shows $x = y$.

Proof

Suppose A has two identity elements, say e and d, relative to the operation $*$. Then $\forall a \in A$, we have $a * e = a = e * a$ and $a * d = a = d * a$. In particular since $d \in A$, $d * e = d = e * d$; and, since $e \in A$, $e * d = e = d * e$. Therefore, $e = e * d = d$. Hence if e is an identity element relative to the operation $*$, then e is unique.

■

Exercises 5.3

1. Refer to Exercises 5.2. For 1-5 determine whether or not the given set has an identity element relative to the defined operation(s). (Be sure to clearly indicate what the identity element is if it exists and then check your result.)

Section 5.4
INVERSE ELEMENTS

In the previous section we noted that the integers possess two special elements: 0 (the additive identity) and 1 (the multiplicative identity). Furthermore, for each integer n there exists another integer m (m = -n) such that n + m = 0 = m + n. However, this is not the case for the operation of multiplication, i.e., if $n \in \mathbf{Z}$ then there does not necessarily exist an integer m for which $n \cdot m = 1 = m \cdot n$. In this section we look at sets having an identity element, e, relative to a given operation, $*$, and attempt to determine which elements of the set when combined under the operation $*$ produce the identity element e. Such elements will be referred to as *inverse* elements.

Definition 5.4.1

Let A have an identity element e relative to the operation $*$ and $a, b \in A$. The element b is called an ***inverse*** of a relative to $*$ if $a * b = e$ and $b * a = e$.

Note that one does not consider the existence or nonexistence of an inverse element relative to an operation $*$ *until* the existence of an identity element relative to the operation $*$ has first been established.

If b is an inverse of a relative to $*$, then a is an inverse of b relative to $*$. Furthermore, since $e \in A$ and $a * e = a = e * a$, $\forall a \in A$, we have $e * e = e = e * e$. Hence, the identity element e is its own inverse relative to $*$.

Example 5.4.1

Refer to Example 5.1.2.

a. In **Z**:

i. 0 is the additive identity element and every integer a has an inverse relative to +; namely, -a, since a + (-a) = 0 and (-a) + a = 0.

ii. 1 is the multiplicative identity element and 1, -1 are the only integers that have an inverse relative to •. To see this consider the following: Let $a \in \mathbf{Z}$. $a \neq 0$. Does there exist $b \in \mathbf{Z}$ such that ab = 1 and ba = 1? Let ab = 1. Then b = 1/a, since $a \neq 0$. If $b \in \mathbf{Z}$, then $a = \pm 1$. Hence, if a = 1, then b = 1. If a = -1, then b = -1. Therefore, 1 and -1 are the only integers that have an inverse relative to •.

b. In P(X):

i. $\emptyset$ is the identity element relative to the operation $\cup$. If $X \neq \emptyset$, then $\exists A \in P(X)$ such that $A \neq \emptyset$. Does there exist $B \in P(X)$ such that $A \cup B = \emptyset$ and $B \cup A = \emptyset$? Now $\forall B \in P(X)$, $A \subseteq A \cup B$. Hence, if $A \neq \emptyset$, then we have $A \cup B \neq \emptyset$, $\forall B \in P(X)$. Thus if $A \neq \emptyset$, then A has no inverse relative to $\cup$.

ii. X is the identity element relative to the operation $\cap$. If $X \neq \emptyset$, then $\exists A \in P(X)$ such that $A \neq X$. Does there exist $B \in P(X)$ such that $A \cap B = X$ and $B \cap A = X$? Now $\forall B \in P(X)$, $A \cap B \subseteq A$. Hence, if $A \neq X$, then we have $A \cap B \neq X$, $\forall B \in P(X)$. Thus if $A \neq X$, then A has no inverse relative to $\cap$.

c. In $F = \{f \mid f{:}A \to A\}$ the identity function **e** is the identity element relative to $\circ$. If $f \in F$, does there exist $g \in F$ such that $f \circ g = \mathbf{e}$ and $g \circ f = \mathbf{e}$? The only elements in F which have an inverse relative to $\circ$ are those functions which are 1–1 correspondences. (See Theorem 3.3.4 and Example 5.1.2, part (c).)

d. In $A = \{x \mid x = 2^m, m \in Z\}$, $e = 2^0$ is the identity element relative to $*$. We wish to determine which elements in A other than the identity (if any) have an inverse relative to $*$. If $x \in A$, does there exist $y \in A$ such that $x * y = e$ and $y * x = e$? Now if $x * y = e$ and if $x = 2^m$ and $y = 2^n$, then we have

$$2^m * 2^n = 2^0$$
$$2^{m+n} = 2^0$$
$$m + n = 0$$
$$n = -m$$

Therefore, the equation $x * y = e$ has a solution in A; i.e., $y = 2^{-m}$. Since $*$ is commutative, $y * x = e$. Hence, if $x = 2^m$, then $y = 2^{-m}$ is an inverse of x relative to $*$.

Theorem 5.4.1

Let $*$ be an operation on A, e the identity element of A relative to $*$ and $a \in A$. If $*$ is associative and if a has an inverse element $b \in A$ relative to $*$, then b is unique.

Proof

Let $*$ be an associative operation on A, e the identity element of A relative to $*$ and $a \in A$. Suppose there exists elements $b, c \in A$ such that b and c are inverses of a relative to $*$. Now $a * b = e = b * a$ and $a * c = e = c * a$. Thus we have

$$\begin{aligned} b &= b * e \\ &= b * (a * c) \\ &= (b * a) * c \\ &= e * c \\ &= c. \end{aligned}$$

Thus, if a has an inverse b relative to $*$, then b is unique.

■

Example 5.4.2

Refer to Example 5.4.1.

a. In **Z** the operation + is associative, 0 is the identity element relative to + and every integer, a, has an unique inverse, -a, relative to +.

b. In $F = \{f \mid f: A \rightarrow A\}$ the operation $\circ$ is associative, the identity function **e** is the identity element relative to $\circ$ and every function, $f \in F$, that is a 1–1 correspondence has an unique inverse, f^{-1}, relative to $\circ$.

In this chapter the concept of a (binary) operation, $*$, on a non-empty set, A, was introduced together with four basic properties: commutative property, associative property, identity element relative to $*$ and the inverse of an element relative to $*$. The following examples summarize these ideas.

Example 5.4.3

Let **Z** be the set of integers; and, for x, y $\in$ **Z**, let $*: \mathbf{Z} \times \mathbf{Z} \rightarrow \mathbf{Z}$ where

$$*: (x,y) \rightarrow x * y = x + y - 1$$

Then $*$ is an operation on **Z**. (Keep in mind that + is a commutative and an associative operation on **Z**).

a. Determine if $*$ is commutative:

$$\begin{aligned} x * y &= x + y - 1 \\ &= y + x - 1 \\ &= y * x \end{aligned}$$

Therefore, $*$ is commutative.

b. Determine if $*$ is associative:

$$\begin{aligned} (x * y) * z &= (x * y) + z - 1 \\ &= (x + y - 1) + z - 1 \\ &= x + y + z - 2 \end{aligned}$$

$$\begin{aligned} x * (y * z) &= x + (y * z) - 1 \\ &= x + (y + z - 1) - 1 \\ &= x + y + z - 2 \end{aligned}$$

Since $(x * y) * z = x * (y * z)$, $*$ is associative.

c. Determine if **Z** has an identity element relative to $*$:

Does there exists an element e $\in$ **Z** such that $e * x = x$ and $x + e = x$ for all x $\in$ **Z**? Let x $\in$ **Z**.

$$\begin{aligned} e * x &= x \\ e + x - 1 &= x \\ e &= 1 \end{aligned}$$

Since 1 $\in$ **Z** and $*$ is commutative, we have $1 * x = x$ and $x * 1 = x$. Hence, 1 is the identity element relative to $*$.

Example 5.4.3 continued next page.

Example 5.4.3 Continued

d. Determine which elements in **Z** have an inverse relative to $*$.

For $x \in \mathbf{Z}$, does there exists $y \in \mathbf{Z}$ such that $y * x = 1$ and $x * y = 1$?
Let $x \in \mathbf{Z}$.

$$\begin{aligned} y * x &= 1 \\ y + x - 1 &= 1 \\ y &= 2 - x \end{aligned}$$

Since $2 - x \in \mathbf{Z}$ and $*$ is commutative, we have $(2 - x) * x = 1$ and $x * (2 - x) = 1$. Hence, if $x \in \mathbf{Z}$, then x has an inverse relative to $*$; i.e., $2 - x$.

Example 5.4.4

As a final example in this chapter, consider the power set, P(X), where $X \neq \emptyset$. For A, B $\in$ P(X), let $+:P(X) \times P(X) \to P(X)$ where

$$+:(A, B) \to A + B = (A - B) \cup (B - A).$$

Since $-$ and $\cup$ are operations on P(X), + is an operation on P(X). This operation is called the *symmetric difference* of A and B. (Note: A + B may also be written in the form $(A \cap B') \cup (B \cap A')$ since, by Theorem 2.3.6, $A - B = A \cap B'$ and $B - A = B \cap A'$.)

a. The following are some properties which govern the set P(X) and the operation +.

   i. + is Commutative:

$$\begin{aligned} A + B &= (A - B) \cup (B - A) \\ &= (B - A) \cup (A - B) \\ &= B + A \end{aligned}$$

   ii. + is Associative:

   See Exercises 5.4, problem (2).

   iii. $\emptyset$ is the identity element of P(X) relative to +:

$$\begin{aligned} A + \emptyset &= (A - \emptyset) \cup (\emptyset - A) \\ &= A \cup \emptyset \\ &= A \end{aligned}$$

   Since + is commutative, $\emptyset + A = A$.

iv. Each element in P(X) is its own inverse relative to +:

$$\begin{aligned} A + A &= (A - A) \cup (A - A) \\ &= \emptyset \cup \emptyset \\ &= \emptyset \end{aligned}$$

b. If we now consider P(X) with the operations + and $\cap$, we have the additional property that $\cap$ is distributive over +. We shall show that $(A \cap B) + (A \cap C) = A \cap (B + C)$ by making use of the results of Theorems 2.3.4, 2.3.5 and 2.3.6.

$(A \cap B) + (A \cap C)$

$= [(A \cap B) - (A \cap C)] \cup [(A \cap C) - (A \cap B)]$

$= [(A \cap B) \cap (A \cap C)'] \cup [(A \cap C) \cap (A \cap B)']$

$= [(A \cap B) \cap (A' \cup C')] \cup [(A \cap C) \cap (A' \cup B')]$

$= [\{(A \cap B) \cap A'\} \cup \{(A \cap B) \cap C'\}] \cup [\{(A \cap C) \cap A'\} \cup \{(A \cap C) \cap B'\}]$

$= [\{A \cap A') \cap B\} \cup \{A \cap (B \cap C')\}] \cup [\{A \cap A') \cap C\} \cup \{A \cap (C \cap B')\}]$

$= [\{\emptyset \cap B\} \cup \{A \cap (B \cap C')\}] \cup [\{\emptyset \cap C\} \cup \{A \cap (C \cap B')\}]$

$= [\emptyset \cup \{A \cap (B \cap C')\}] \cup [\emptyset \cup \{A \cap (C \cap B')\}]$

$= [A \cap (B \cap C')] \cup [A \cap (C \cap B')]$

$= A \cap [(B \cap C') \cup (C \cap B')]$

$= A \cap [(B - C) \cup (C - B)]$

$= A \cap (B + C)$

Since $\cap$ is a commutative operation on P(X), by Theorem 5.2.1, $\cap$ is distributive over +.

Example 5.4.4 continued next page.

Example 5.4.4 Continued

c. We may summarize the properties of P(X) relative to the operations + and $\cap$ as follows: For A, B, C $\in$ P(X),

$< P(X); + >$

$A + B = B + A$	(Commutative law)
$A + (B + C) = (A + B) + C$	(Associative law)
$A + \emptyset = A$	($\emptyset$ is identity element)
$A + A = \emptyset$	(A is its own inverse)

$< P(X); \cap >$

$A \cap B = B \cap A$	(Commutative law)
$A \cap (B \cap C) = (A \cap B) \cap C$	(Associative law)
$A \cap X = A$	(X is identity element)
$A \cap A = A$	(Idempotent law)

$< P(X); +, \cap >$

$A \cap (B + C) = (A \cap B) + (A \cap C)$	(Distributive law)

d. P(X) together with the operations of + and $\cap$ is an example of a mathematical system called a *Boolean ring.*

e. As a concrete example of a Boolean Ring, consider X = {a,b} and P(X) = {$\emptyset$,{a},{b},X}. For the operations of + and $\cap$, we have the following tables:

+	$\emptyset$	{a}	{b}	X
$\emptyset$	$\emptyset$	{a}	{b}	X
{a}	{a}	$\emptyset$	X	{b}
{b}	{b}	X	$\emptyset$	{a}
X	X	{b}	{a}	$\emptyset$

$\cap$	$\emptyset$	{a}	{b}	X
$\emptyset$	$\emptyset$	$\emptyset$	$\emptyset$	$\emptyset$
{a}	$\emptyset$	{a}	$\emptyset$	{a}
{b}	$\emptyset$	$\emptyset$	{b}	{b}
X	$\emptyset$	{a}	{b}	X

Exercises 5.4

1. Refer to Exercises 5.2. For 1-5, if the given set has an identity element relative to the defined operations, determine which elements in the set other than the identity (if any) have an inverse. (Be sure to clearly indicate what the inverse element is if it exists and then check your result.)

2. Show that the operation + as defined on P(X) in Example 5.4.4 is an associative operation. (Hint: Make use of the results of Theorems 2.3.4, 2.3.5 and 2.3.6. It may be helpful to calculate A + (B + C) and (A + B) + C separately to a point where it can be shown that equality holds.)

3. Let A = {a,b,c,d} and $*$ be the operation defined on A by the given table.

$*$	a	b	c	d
a	c	a	d	b
b	a	b	c	d
c	d	c	b	a
d	b	d	a	c

   a. Is the operation $*$ commutative?

   b. Does A have an identity element relative to $*$?

   c. If A has an identity element relative to $*$, which elements of A have inverses relative to $*$?

4. Refer to Example 5.4.4. Let A be a *fixed* element of the power set P(X) and consider the function f: P(X) → P(X) where $\forall B \in P(X)$, f(B) = A + B + A.

   a. Show f is a 1–1 correspondence.

   b. Show f (B + C) = f(B) + f(C) $\forall B, C \in P(X)$.

Chapter 6

NUMBER THEORY

Number theory is concerned with properties of the integers. You are already familiar with the arithmetic of the integers and many of the elementary properties of the integers. In this chapter we shall establish some additional basic concepts in number theory. The fascination with number theory can be traced back to very early mathematics. In fact, some of the properties examined in this chapter are in much the same form as they were then.

Section 6.1
THE INTEGERS AND DIVISIBILITY

Recall that $\mathbf{Z} = \{0,\pm1,\pm2,\pm3,\ldots\}$ is the set of integers and that $\mathbf{Z}$ is closed under $+$ and $\bullet$; i.e., whenever $a, b \in \mathbf{Z}$, then $a + b$ and $a \bullet b$ are unique integers. Let us review some of the properties previously discussed that govern the integers, $\mathbf{Z}$, relative to the operations $+$ and $\bullet$.

Addition

$a + b = b + a$	$\forall a, b \in \mathbf{Z}$	(commutative law)
$a + (b + c) = (a + b) + c$	$\forall a, b, c \in \mathbf{Z}$	(associative law)
$a + 0 = a$	$\forall a \in \mathbf{Z}$	(additive identity)
$\forall a \in \mathbf{Z}\ \exists(-a) \in \mathbf{Z}$ such that	$a + (-a) = 0$	(additive inverse)

Multiplication

$ab = ba$	$\forall a, b \in \mathbf{Z}$	(commutative law)
$a(bc) = (ab)c$	$\forall a, b, c \in \mathbf{Z}$	(associative law)
$a \bullet 1 = a$	$\forall a \in \mathbf{Z}$	(multiplicative identity)

Addition and Multiplication

$a(b + c) = ab + ac$	$\forall a, b, c \in \mathbf{Z}$	(distributive law)

Throughout this chapter and subsequent chapters it will be convenient to use the following letters to denote some additional common sets.

$\mathbf{Z}^+$ = set of positive integers = $\{1,2,3,\ldots\}$

$\mathbf{Z}^-$ = set of negative integers = $\{-1,-2,-3,\ldots\}$

$\mathbf{E}$ = set of even integers = $\{0,\pm2,\pm4,\ldots\} = \{x \mid x = 2n, n \in \mathbf{Z}\}$

$\mathbf{O}$ = set of odd integers = $\{\pm1,\pm3,\pm5,\ldots\} = \{x \mid x = 2n - 1, n \in \mathbf{Z}\}$

The next theorem is a very important and necessary result needed in the development of number theory. Simply stated, it says every non-empty set of positive integers contains a least or smallest element.

Theorem 6.1.1 (Well-Ordering Principle)

Let $A \subseteq \mathbf{Z}^+$, $A \neq \emptyset$. Then A has a smallest element; i.e., there exists $m \in A$ such that $\forall a \in A$, $m \leq a$.

Proof

Let $A \subseteq \mathbf{Z}^+$, $A \neq \emptyset$. We have two cases to consider.

a. $\mathbf{1 \in A}$: Since $A \subseteq \mathbf{Z}^+$ and $\forall x \in \mathbf{Z}^+$, $1 \leq x$, we have $\forall a \in A$, $1 \leq a$. Thus A has a smallest element; i.e., 1.

b. $\mathbf{1 \notin A}$: Assume A does not have a smallest element. Let $B = \mathbf{Z}^+ - A = \{x \in \mathbf{Z}^+ \mid x \notin A\}$; and, for each $n \in \mathbf{N}$, let p(n) be the statement: $\{1,2,\ldots,n\} \subseteq B$.

p(1): Show p(1) is true; i.e., $\{1\} \subseteq B$. Since $1 \notin A$, $1 \in B$. Hence, $\{1\} \subseteq B$. Therefore p(1) is true.

p(k): Assume p(k) is true for some k, $k \geq 1$. Then $\{1,2,\ldots,k\} \subseteq B$.

p(k + 1): Show p(k + 1) is true: i.e., $\{1,2,\ldots,k,k+1\} \subseteq B$. Since $\{1,2,\ldots,k\} \subseteq B$ and $B = \mathbf{Z}^+ - A$, then $1,2,\ldots,k \notin A$. Thus if $a \in A$, $k + 1 \leq a$. But A has no smallest element and hence $\forall a \in A$, $k + 1 < a$. Therefore, $k + 1 \notin A$ which implies $k + 1 \in B$ and we have $\{1,2,\ldots,k,k+1\} \subseteq B$. Therefore, p(k + 1) is true.

Hence $(\forall n)(p(n))$ is true. Consequently, if $n \in \mathbf{Z}^+$, then $n \in \{1,2,\ldots,n\} \subseteq B$ and we have $\mathbf{Z}^+ \subseteq B$. Since $B \subseteq \mathbf{Z}^+$, $B = \mathbf{Z}^+$. Thus $A = \emptyset$. Hence, we have $(A \neq \emptyset) \wedge (A = \emptyset)$ which is a contradiction that arose from assuming A did not have a smallest element. Therefore, A has a smallest element.

■

Theorem 6.1.2

Let $a, b \in \mathbf{Z}$.

a. If $a, b \in \mathbf{E}$, then $a + b \in \mathbf{E}$.

b. If $a, b \in \mathbf{O}$, then $a + b \in \mathbf{E}$.

c. If $a \in \mathbf{E}$ and $b \in \mathbf{O}$, then $a + b \in \mathbf{O}$.

Proof

We shall prove part (b) and leave the proofs of parts (a) and (c) as an exercise.

b. Let a, b $\in$ **O**. Then $\exists$h, k $\in$ **Z** such that

$$a = 2h - 1$$
$$b = 2k - 1$$

Therefore, $a + b = (2h - 1) + (2k - 1)$
$= 2h + 2k - 2$
$= 2(h + k - 1)$
$= 2m$ where $m = h + k - 1 \in \mathbf{Z}$

Hence, $a + b \in \mathbf{E}$. ■

Definition 6.1.1

Let a, d $\in$ **Z**, d $\neq$ 0. Then d ***divides*** a and we write **d | a** if there exists c $\in$ **Z** such that a = dc.

There are other ways to express the concept of "d divides a." We could say "d is a *divisor* of a," "d is a *factor* of a" or "a is a *multiple* of d."

Example 6.1.1

a. $3 \mid 12$ since $4 \in \mathbf{Z}$ and $12 = 3 \cdot 4$.

b. $-3 \mid 12$ since $-4 \in \mathbf{Z}$ and $12 = (-3) \cdot (-4)$.

c. 5 does not divide 12, denoted by $5 \nmid 12$, since there does not exist $c \in \mathbf{Z}$ such that $12 = 5 \cdot c$.

d. If $a \in \mathbf{Z}$, $a \neq 0$, then $a \mid 0$ since $0 \in \mathbf{Z}$ and $0 = a \cdot 0$.

Theorem 6.1.3

Let a, b, d $\in$ **Z**.

a. If d | a, then d | (-a). [In particular, if d | a, then d | (| a |).]

b. If d | a, then -d | a. [In particular, if d | a, then (| d |) | a.]

c. If d | a and a $\neq$ 0, then d $\leq$ | a |.

d. If a | b and b | a, then a = $\pm$b.

e. If d | a and k $\in$ **Z**, k $\neq$ 0, then kd | ka.

f. If d | a and a | b, then d | b.

g. If d | a and d | b, then for any r, s $\in$ **Z**, d | (ra + sb).
[In particular, if d | a and d | b, then d | (a + b), d | ra and d | sb.]

Proof

We shall prove parts (a), (c), (e) and (g) and leave the proofs of parts (b), (d) and (f) as an exercise.

Let a, b, d $\in$ **Z**

a. Let d | a. Then $\exists$c $\in$ **Z** such that a = dc. Therefore, -a = -(dc) = d(-c). Since -c $\in$ **Z**, d | (-a).

■

c. Let d | a and a $\neq$ 0. Since d | a, d | (-a). Hence d | (| a |). Thus $\exists$c $\in$ **Z** such that | a | = dc. Since a $\neq$ 0, | a | > 0. Therefore, either d < 0 and c < 0 or d > 0 and c > 0.

   i. **d < 0 and c < 0**: Then we have d < 0 and 0 < | a | which yields d < | a |.

   ii. **d > 0 and c > 0**: Since c > 0 and c $\in$ **Z**, 1 $\leq$ c. Thus d $\leq$ dc. But | a | = dc. Therefore, d $\leq$ | a |.

■

e. Let d | a and k $\in$ **Z**, k $\neq$ 0. Since d | a, $\exists$c $\in$ Z such that a = dc. Thus ka = k(dc) = (kd)c; and, since kd $\neq$ 0, kd | ka.

■

Proof continued next page.

g. Let $d \mid a$ and $d \mid b$. Thus $\exists m, n \in \mathbf{Z}$ such that $a = dm$ and $b = dn$. Let $r, s \in \mathbf{Z}$. Then

$$\begin{aligned} ra + sb &= r(dm) + s(dn) \\ &= d(rm) + d(sn) \\ &= d(rm + sn) \\ &= dc \text{ where } c = rm + sn \in \mathbf{Z}. \end{aligned}$$

Hence, $d \mid (ra + sb)$. (In particular, if $r = s = 1$, $d \mid (a + b)$; and, if $s = 0$ or $r = 0$, we have $d \mid ra$ or $d \mid sb$.)

■

Given two integers, we know that one integer does not necessarily divide the other. The following theorem allows us to generalize the concept of divide to include what is commonly referred to as the quotient and remainder.

Theorem 6.1.4 (Division Algorithm)

Let $a, b \in \mathbf{Z}$, $a \neq 0$. Then there exist unique $q, r \in \mathbf{Z}$ such that $b = aq + r$ and $0 \leq r < |a|$.

Discussion
In addition to showing uniqueness, which we have already discussed, the proof of this theorem also involves a statement of the form $p \Rightarrow (q \wedge r)$. Recall from Exercises 1.2, (9), that $[p \Rightarrow (q \wedge r)] \equiv [(p \Rightarrow q) \wedge (p \Rightarrow r)]$ and that $(p \Rightarrow q) \wedge (p \Rightarrow r)$ is true only if $p \Rightarrow q$ and $p \Rightarrow r$ are both true. Hence, to show

$$[(a, b \in \mathbf{Z}) \wedge (a \neq 0)] \Rightarrow (\exists q, r \in \mathbf{Z})[(b = aq + r) \wedge (0 \leq r < |a|)]$$

is true, we must show that for $a, b \in \mathbf{Z}$ and $a \neq 0$, we can find $q, r \in \mathbf{Z}$ such that (1) $b = aq + r$ is true and (2) $0 \leq r < |a|$ is true.

Proof

Let $a, b \in \mathbf{Z}$, $a \neq 0$.

1. We first show that for $a > 0$ and any $b \in \mathbf{Z}$, $\exists q, r \in \mathbf{Z}$ such that $b = aq + r$ and $0 \leq r < a$. We have two cases to consider.

   i. **$a \mid b$**: Then $\exists q \in \mathbf{Z}$ such that $b = aq = aq + 0$
   Hence, $r = 0$.

   ii. **$a \nmid b$**: Then $a \neq 1$ and $b \neq 0$. Thus we have $a \geq 2$ and $b > 0$ or $b < 0$.
   Let $A = \{b - ax \mid x \in \mathbf{Z} \text{ and } b - ax > 0\}$. Now $A \neq \emptyset$.

   $b > 0$: Then for $x = -1$, we have $b - a(-1) = b + a > 0$.
   Hence $b - a(-1) \in A$.

   $b < 0$: Then for $x = b$, we have $b - ab = b - ba = b(1 - a)$.
   Since $a \geq 2$, $-a \leq -2$ and thus $1 - a \leq -1 < 0$. Hence, $b < 0$ and $1 - a < 0$ gives us $b(1 - a) > 0$. Therefore $b - ab \in A$.

   Since $A \subseteq \mathbf{Z}^+$ and $A \neq \emptyset$, by Theorem 6.1.1 (Well-Ordering Principle), A has a smallest element, say r. But $r \in A$ implies $r = b - aq$ where $q \in \mathbf{Z}$ and $b - aq > 0$. Thus $b = aq + r$ and $r > 0$.

   We need only show $r < a$. Suppose $r \geq a$.

   $r = a$: Since $r = b - aq$, $a = b - aq$ or $b = a(q + 1)$.
   Thus $a \mid b$. Contradiction.

   $r > a$: Then $r - a > 0$; and,
   since $r - a = (b - aq) - a = b - a(q + 1)$, $r - a \in A$. But $a > 0$ implies $r - a < r$. Thus r is not the smallest element of A.
   Contradiction.

   Hence, we have $r < a$ and thus, when $a \nmid b$, $0 < r < a$.

   Now we have shown that for $a > 0$, $b = aq + r$ and $0 \leq r < a$. Since $a = (-a)(-1)$, $b = [(-a)(-1)q + r] = (-a)(-q) + r$. Since $a \mid (-a)$ and $-a \neq 0$, $a \leq |-a|$ by Theorem 6.1.3, part (c). Thus $0 \leq r < a \leq |-a|$ or $0 \leq r < |-a|$.
   Therefore, the theorem is true for any $a, b \in \mathbf{Z}$, $a \neq 0$.

Proof continued next page.

2. To complete the proof of the theorem we must show that q and r are unique. Suppose $b = aq + r$ and $b = aq' + r'$ where $0 \le r, r' < |a|$. We may assume that $r \le r'$. Then

$$\begin{aligned} aq + r &= aq' + r' \\ aq - aq' &= r' - r \\ a(q - q') &= r' - r \\ |a(q - q'| &= |r' - r| \end{aligned}$$

Now $r \le r'$ implies $0 \le r' - r$ and thus $|r' - r| = r' - r$. Since $r < |a|$ and $r' < |a|$, $r' - r < |a|$. Thus we have

$$\begin{aligned} |a(q - q')| &= r' - r \\ |a(q - q')| &< |a| \\ |a| \bullet |q - q'| &< |a| \\ |q - q'| &< 1 \end{aligned}$$

Since $|q - q'| \in \mathbf{Z}$ and $0 \le |q - q'| < 1$, $|q - q'| = 0$. Hence, $q - q' = 0$ or $q = q'$. This yields $r' - r = 0$ or $r = r'$. Therefore, q and r are unique. ■

Example 6.1.1

a. If $a = 12$, then for any $b \in \mathbf{Z}$, $b = 12q + r$ and $0 \le r < |12|$ or $r = 0, 1, 2, \ldots, 11$. Hence if $b = 75$, then $75 = 12 \bullet 6 + 3$ and $0 < 3 < |12|$.

b. If $a = 4$, then for any $b \in \mathbf{Z}$, $b = 4q + r$ and $0 \le r < |4|$ or $r = 0, 1, 2, 3$. Hence if $b = -63$, then $-63 = 4(-16) + 1$ and $0 < 1 < |4|$.

c. If $a = -4$, then for any $b \in \mathbf{Z}$, $b = (-4)q + r$ and $0 \le r < |-4|$ or $r = 0, 1, 2, 3$. Hence, if $b = 11$, then $11 = (-4)(-2) + 3$ and $0 < 3 < |-4|$.

d. If $a = -5$, then for any $b \in \mathbf{Z}$, $b = (-5)q + r$ and $0 \le r < |-5|$ or $r = 0, 1, 2, 3, 4$. Hence, if $b = -17$, then $-17 = (-5)(4) + 3$ and $0 < 3 < |-5|$.

Exercises 6.1

1. Prove parts (a) and (c) of Theorem 6.1.2.

2. Let a, b $\in$ **Z**. Prove the following.

   a. If a, b $\in$ **E**, then ab $\in$ **E**.

   b. If a, b $\in$ **O**, then ab $\in$ **O**.

   c. If a $\in$ **E** and b $\in$ **O**, then ab $\in$ **E**.

3. Prove parts (b), (d) and (f) of Theorem 6.1.3.

4. For a, b $\in$ **Z**, let $a \mathcal{R} b \Leftrightarrow a \mid b$. Is the relation reflexive?, symmetric?, transitive?

5. Find q and r satisfying the division algorithm (Theorem 6.1.4) for

   a. a = 33, b = 826

   b. a = -16, b = 780

6. Let a be a fixed positive integer, $a \geq 2$. Show that the function f: **N** $\rightarrow$ **N**, where $f(n) = a^n$, $\forall n \in$ **N**, is 1–1.

Section 6.2

GREATEST COMMON DIVISOR AND LEAST COMMON MULTIPLE

Suppose we are given a pair of integers, say a and b. There are two situations we wish to address in this section. First, we want to find the common divisors (factors) of a and b; and, if such divisors exist, then determine if there is a "largest" divisor of a and b. Secondly, we want to find all those integers which have a and b as divisors (factors) and, if such integers exist, then determine if there is a "smallest" such integer that has a and b as divisors. We shall see that these two situations are related.

Definition 6.2.1

Let a, b, $c \in \mathbf{Z}$, $c > 0$. Then c is called a ***greatest common divisor (gcd)*** of a and b and we write $\mathbf{c = (a, b)}$ if:

a. $c \mid a$ and $c \mid b$.

b. whenever $x \mid a$ and $x \mid b$, where $x \in \mathbf{Z}$, then $x \mid c$.

It should be noted that if c = (a,b), then not only does c divide both a and b, but c is the largest positive integer that divides both a and b.

Example 6.2.1

a. Show 6 = (12,18).

   i. $6 \mid 12$ since $12 = 6 \bullet 2$ and $6 \mid 18$ since $18 = 6 \bullet 3$.

   ii. Let $x \mid 12$ and $x \mid 18$. Since $x \mid 12$, then $x \in \{\pm1,\pm2,\pm3,\pm4,\pm6,\pm12\}$. Since $x \mid 18$, $x \in \{\pm1,\pm2,\pm3,\pm6,\pm9,\pm18\}$. Hence, if $x \mid 12$ and $x \mid 18$, then

   $$x \in \{\pm1,\pm2,\pm3,\pm4,\pm6,\pm12\} \cap \{\pm1,\pm2,\pm3,\pm6,\pm9,\pm18\} = \{\pm1,\pm2,\pm3,\pm6\}.$$

   Thus $x \mid 6$.

b. Show 2 = (-4,6)

   i. $2 \mid (-4)$ since $-4 = 2 \bullet (-2)$ and $2 \mid 6$ since $6 = 2 \bullet 3$.

   ii. Let $x \mid (-4)$ and $x \mid 6$. Since $x \mid (-4)$, $x \in \{\pm1,\pm2,\pm4\}$. Since $x \mid 6$, then $x \in \{\pm1,\pm2,\pm3,\pm6\}$. Hence, if $x \mid (-4)$ and $x \mid 6$, then

   $$x \in \{\pm1,\pm2,\pm4\} \cap \{\pm1,\pm2,\pm3,\pm6\} = \{\pm1,\pm2\}.$$

   Thus $x \mid 2$.

Theorem 6.2.1

Let $a, b, c \in \mathbf{Z}$.

a. If $c = (a,b)$, then c is unique.

b. If $c = (a,b)$, then $c = (|a|,|b|)$. [In particular, if $c = (a,b)$, then $c = (-a,b) = (a,-b) = (-a,-b)$.]

c. If $a \in \mathbf{Z}$, $a \neq 0$, then $|a| = (a,0)$.

d. The pair 0 and 0 has no gcd.

Proof

Let $a, b, c \in \mathbf{Z}$. We shall prove parts (a) and (d) and leave the proofs of parts (b) and (c) as exercises.

a. Suppose $c = (a,b)$ and $d = (a,b)$. Since $d \mid a$, $d \mid b$ and $c = (a,b)$, $d \mid c$. Since $c \mid a$, $c \mid b$ and $d = (a,b)$, $c \mid d$. By Theorem 6.1.3, part (d), $d = \pm c$. Since $d > 0$ and $c > 0$, $d = c$. Therefore, if $c = (a,b)$, then c is unique.

■

d. Suppose $c = (0,0)$. Since $c \in \mathbf{Z}$ and $c > 0$, $c \geq 1$ which implies $c + 1 \geq 2$. Therefore, since $c + 1 \neq 0$, $(c + 1) \mid 0$ (see Example 6.1.1, part (d)). But $c = (0,0)$ implies that $(c + 1) \mid c$. Hence, we have $((c + 1) \nmid c) \wedge ((c + 1) \mid c)$ which is a contradiction. Thus the pair 0 and 0 has no gcd.

■

For fixed integers a and b, the integer equal to $au + bv$ for any $u, v \in \mathbf{Z}$ is called a *linear combination* of a and b. The following theorem states that the greatest common divisor of a and b, if it exists, is some linear combination of a and b.

Theorem 6.2.2

Let $a, b \in \mathbf{Z}$, $a \neq 0$ and $b \neq 0$. Then a and b have a gcd, c, and there exist $u, v \in \mathbf{Z}$ such that $c = (a, b) = au + bv$.

Proof

Let $a, b \in \mathbf{Z}$, $a \neq 0$ and $b \neq 0$. By Theorem 6.2.1, part (b), it suffices to show the existence of c for $a > 0$, $b > 0$.

Let $A = \{ah + bk \mid h, k \in \mathbf{Z} \text{ and } ah + bk > 0\}$. If $h = 1$ and $k = 0$, then $ah + bk = a \cdot 1 + b \cdot 0 = a > 0$. Therefore, $A \neq \emptyset$. Now $A \subseteq \mathbf{Z}^+$ and $\mathbf{Z}^+$ is well-ordered. Hence A has a smallest element, say c. Since $c \in A$, $\exists u, v \in \mathbf{Z}$ such that $c = au + bv$ and $c = au + bv > 0$. We shall show $c = (a,b)$.

1. $c \mid a$ and $c \mid b$: Applying the division algorithm to c and a, $\exists$ unique $q, r \in \mathbf{Z}$ such that $a = cq + r$ and $0 \leq r < |c| = c$.

   Therefore,
   $$\begin{aligned} a &= qc + r \\ &= q(au + bv) + r \\ &= (auq) + (bvq) + r \end{aligned}$$
   So,
   $$r = a - (auq) - (bvq) = a(1 - uq) + b(-vq).$$

Let $x = 1 - uq$ and $y = -vq$. Then $x, y \in \mathbf{Z}$ and $r = ax + by$. Suppose $r \neq 0$. Then $r = ax + by > 0$ and thus $r \in A$. But $0 < r < c$ which contradicts the choice of c. Hence, $r = 0$ and $a = cq$ or $c \mid a$. In like manner one can apply the division algorithm to c and b and show $c \mid b$.

2. Let $x \mid a$ and $x \mid b$. We must show $x \mid c$. Since $x \mid a$ and $x \mid b$, by Theorem 6.1.3, part (g), $x \mid (au + bv)$. But $c = au + bv$. Therefore, $x \mid c$.

Hence, $c = (a,b)$ and $c = au + bv$.

■

Theorem 6.2.3

Let $a, b \in \mathbf{Z}$ and $c = (a,b)$. If $k \in \mathbf{Z}$, $k \neq 0$, then $|k|c = (ka,kb)$.

Proof

Let $a, b \in \mathbf{Z}$, $c = (a,b)$ and $k \in \mathbf{Z}$ such that $k \neq 0$. Since $|k|c > 0$, we must show:

$(|k|c) \mid (ka)$ and $(|k|c) \mid (kb)$
If $x \mid (ka)$ and $x \mid (kb)$, then $x \mid (|k|c)$.

Now $c = (a,b)$ implies that $c \mid a$ and $c \mid b$. Since $k \neq 0$, $(kc) \mid (ka)$ and $(kc) \mid (kb)$ by Theorem 6.1.3, part (e). Hence, by Theorem 6.1.3, part (b), $(|kc|) \mid (ka)$ and $(|kc|) \mid (kb)$.

But $|kc| = |k| \cdot |c| = |k|c$. Thus we have $(|k|c) \mid (ka)$ and $(|k|c) \mid (kb)$.

Let $x \mid (ka)$ and $x \mid (kb)$. Since $c = (a,b)$, $\exists u, v \in \mathbf{Z}$ such that $c = au + bv$. By Theorem 6.1.3, part (g), since $x \mid (ka)$, $x \mid (kb)$ and $u, v \in \mathbf{Z}$, $x \mid [u(ka + v(kb)]$. But $u(ka) + v(kb) = k(au + bv) = kc$. Hence, $x \mid (kc)$ and thus $x \mid (|kc|)$. But $|kc| = |k|c$. Thus we have $x \mid (|k|c)$.

Therefore, $|k|c = (ka,kb)$.

■

Note that, if $a, b \in \mathbf{Z}$ where $a \neq 0$ or $b \neq 0$, then by Theorem 6.2.1, part (c), and Theorem 6.2.2, a and b have a gcd, c. To find c, find the largest positive integer that divides both a and b.

Example 6.2.2

Consider the pair of integers 8 and -12.

a. We wish to find the gcd, c, for 8 and -12 and express c in the form $c = 8u + (-12)v$ where $u, v \in \mathbf{Z}$.

   Now 4 is the largest positive integer such that $4 \mid 8$ and $4 \mid (-12)$. Thus we have $4 = (8,-12)$.

   Since $4 = (8,-12)$, $\exists u, v \in \mathbf{Z}$ such that $4 = 8u + (-12)v$. Now $4 = 8 \cdot (-1) + (-12) \cdot (-1)$. Also, $4 = 8 \cdot 2 + (-12) \cdot 1$. Thus we see that u and v are not unique.

b. Now $4 = (8,-12)$. From Theorem 6.2.3, if we let $k = -3$, then $|-3| \cdot 4 = ((-3)\,8, (-3)(-12))$ or $12 = (-24,36)$.

From Theorem 6.2.1, part (b), we know that $(a,b) = (|a|,|b|)$. The next theorem introduces a technique called the *Euclidean Algorithm* which enables one to calculate the gcd by repeated application of the division algorithm (Theorem 6.1.4). Once the gcd of a and b has been found, the Euclidean Algorithm also gives us a way to express (a,b) as a linear combination of a and b by reversing our steps.

Theorem 6.2.4 (Euclidean Algorithm)

Let a, b $\in \mathbf{Z}^+$, where $a \le b$. Thus by the division algorithm:

$$
\begin{array}{ll}
b = aq_1 + r_1 & 0 \le r_1 < a \\
a = r_1 q_2 + r_2 & 0 \le r_2 < r_1 \\
r_1 = r_2 q_3 + r_3 & 0 \le r_3 < r_2 \\
\bullet & \bullet \\
\bullet & \bullet \\
\bullet & \bullet \\
r_{k-3} = r_{k-2} q_{k-1} + r_{k-1} & 0 \le r_{k-1} < r_{k-2} \\
r_{k-2} = r_{k-1} q_k + r_k & 0 \le r_k < r_{k-1} \\
r_{k-1} = r_k q_{k+1} + 0 &
\end{array}
$$

Then $r_k = (a,b)$.

Proof

Let a, b $\in \mathbf{Z}^+$, where $a \le b$ and $c = (a,b)$.

First note that $r_1 > r_2 > r_3 \ldots$; i.e., the repeated application of the division algorithm results in smaller remainders. Since each $r_i \in \mathbf{Z}$ and $r_i \ge 0$, one of the remainders, say r_{k+1}, must eventually be 0. If not, then we would have a set of positive integers, $\{r_1, r_2, r_3, \ldots, r_k, \ldots\}$, having no smallest element. This would contradict the Well-Ordering Principle.

We shall show $c \mid r_k$ and $r_k \mid c$ and thus $c = r_k$.

1. **$c \mid r_k$**: Now $r_1 = b - aq_1 = 1 \cdot b + (-q_1)a$. Since $c = (a,b)$, $c \mid a$ and $c \mid b$ and thus by Theorem 6.1.3, part (g), $c \mid (1 \cdot b + (-q_1)a)$. Therefore, $c \mid r_1$.

   Now $r_2 = a - r_1 q_2 = 1 \cdot a + (-q_2)r_1$. Since $c \mid a$ and $c \mid r_1$, $c \mid (1 \cdot a + (-q_2)r_1)$. Therefore, $c \mid r_2$.

   Continuing the process, we have $c \mid r_3, \ldots, c \mid r_k$.

2. $r_k \mid c$: Now $r_{k-1} = r_k q_{k+1}$ which implies $r_k \mid r_{k-1}$.
Since $r_k \mid r_{k-1}$ and $r_k \mid r_k$, $r_k \mid (q_k \cdot r_{k-1} + 1 \cdot r_k)$ by Theorem 6.1.3, part (g).
But $r_{k-2} = r_{k-1}q_k + r_k$. Therefore, $r_k \mid r_{k-2}$.

Since $r_k \mid r_{k-2}$ and $r_k \mid r_{k-1}$, $r_k \mid (q_{k-1} \cdot r_{k-2} + 1 \cdot r_{k-1})$.
But $r_{k-3} = r_{k-2}q_{k-1} + r_{k-1}$. Hence $r_k \mid r_{k-3}$.

Continuing the process, we have $r_k \mid r_{k-4}, \ldots, r_k \mid r_1, r_k \mid a, r_k \mid b$. Since $r_k \mid a$ and $r_k \mid b$ and $c = (a,b)$, $r_k \mid c$.

Now we have $c \mid r_k$ and $r_k \mid c$ which implies $c = \pm r_k$. But $c > 0$ and $r_k > 0$. Therefore, $c = r_k$ and hence $r_k = (a,b)$.

■

Example 6.2.3

a. Consider the pair of integers 48 and 68. Use the Euclidean Algorithm to find the gcd, c, for 48 and 68 and express c in the form $c = 48u + 68v$, where $u, v \in \mathbf{Z}$.

Determine (48, 68).

$$68 = 48 \cdot 1 + 20$$
$$48 = 20 \cdot 2 + 8$$
$$20 = 8 \cdot 2 + 4$$
$$8 = 4 \cdot 2 + 0$$

Hence, $(48,68) = 4$

Determine $u, v \in \mathbf{Z}$ such that $4 = (48,68) = 48u + 68v$.

$$\begin{aligned}
4 &= 20 - 8 \cdot 2 \\
&= 20 - (48 - 20 \cdot 2) \cdot 2 \\
&= 20 - 48 \cdot 2 + 20 \cdot 4 \\
&= 20 \cdot 5 - 48 \cdot 2 \\
&= (68 - 48 \cdot 1) \cdot 5 - 48 \cdot 2 \\
&= 68 \cdot 5 - 48 \cdot 5 - 48 \cdot 2 \\
&= 68 \cdot 5 - 48 \cdot 7 \\
&= 68 \cdot 5 + 48(-7) \\
&= 48 \cdot (-7) + 68 \cdot 5
\end{aligned}$$

Hence, $u = -7$ and $v = 5$.

Example 6.2.3 continued next page.

Example 6.2.3 Continued

b. Consider the pair of integers -24 and 63. Use the Euclidean Algorithm to find the gcd, c, for -24 and 63 and express c in the form c = (-24)u + 63v.

Determine (-24, 63). Since (-24,63) = (24,63), use the Euclidean Algorithm to find the gcd of 24 and 63.

$$\begin{aligned} 63 &= 24 \bullet 2 + 15 \\ 24 &= 15 \bullet 1 + 9 \\ 15 &= 9 \bullet 1 + 6 \\ 9 &= 6 \bullet 1 + 3 \\ 6 &= 3 \bullet 2 + 0 \end{aligned}$$

Hence, (-24,63) = (24,63) = 3.

Determine u, v $\in$ **Z** such that 3 = (-24,63) = (-24)u + 63v. Use the above result to first find u, v $\in$ **Z** for the pair 24 and 63 such that 3 = 24u + 63v.

$$\begin{aligned} 3 &= 9 - 6 \bullet 1 \\ &= 9 - (15 - 9 \bullet 1) \bullet 1 \\ &= 9 - 15 \bullet 1 + 9 \bullet 1 \\ &= 9 \bullet 2 - 15 \bullet 1 \\ &= (24 - 15 \bullet 1) \bullet 2 - 15 \bullet 1 \\ &= 24 \bullet 2 - 15 \bullet 2 - 15 \bullet 1 \\ &= 24 \bullet 2 - 15 \bullet 3 \\ &= 24 \bullet 2 - (63 - 24 \bullet 2) \bullet 3 \\ &= 24 \bullet 2 - 63 \bullet 3 + 24 \bullet 6 \\ &= 24 \bullet 8 - 63 \bullet 3 \\ &= 24 \bullet 8 + 63 \bullet (-3) \end{aligned}$$

Now for the pair -24 and 63 we have 3 = (-24)(-8) + 63(-3).
Hence, u = -8, and v = -3.

Definition 6.2.2

Let a, b $\in$ **Z**. We say that a and b are ***relatively prime*** if 1 = (a,b).

Note that two integers are *relatively prime* if their only positive common divisor is 1.

Example 6.2.4

Determine whether the pair of integers 5 and -12 is relatively prime.

Since (5,-12) = (5,12), we shall consider the pair 5 and 12 and use the Euclidean Algorithm to find the gcd.

$$\begin{aligned} 12 &= 5 \bullet 2 + 2 \\ 5 &= 2 \bullet 2 + 1 \\ 2 &= 2 \bullet 1 + 0 \end{aligned}$$

Since 1 = (5,12), we have 1 = (5,-12) and thus the integers 5 and -12 are relatively prime.

Theorem 6.2.5

Let a, b, c $\in$ **Z**.

a. If a | bc and 1 = (a,b), then a | c.

b. If c = (a,b) and if $a = a_1c$ and $b = b_1c$, then $1 = (a_1,b_1)$.

Proof

Let a, b, c $\in$ **Z**.

a. Let 1 = (a,b) and a | bc. Since 1 = (a,b), $\exists$u, v $\in$ **Z** such that 1 = au + bv. Hence, c • 1 = c(au + bv) or c = a(cu) + (bc)v. Now a | bc implies $\exists$w $\in$ **Z** such that bc = aw. Thus we have

$$\begin{aligned} c &= a(cu) + (bc)v \\ &= a(cu) + (aw)v \\ &= a(cu) + a(wv) \\ &= a(cu + wv) \\ &= ad \text{ where } d = cu + wv \in \mathbf{Z}. \end{aligned}$$

Therefore, a | c.

■

b. Let c = (a,b), $a = a_1c$ and $b = b_1c$.
Now $c = (a,b) = (a_1c,b_1c) = |c|(a_1,b_1) = c \bullet (a_1,b_1)$ by Theorem 6.2.3.
Thus we have $c \bullet 1 = c \bullet (a_1,b_1)$ or $1 = (a_1,b_1)$.

■

Example 6.2.5

a. Since $5 \mid (-12 \cdot 10)$ and $1 = (5,-12)$, by Theorem 6.2.5, $5 \mid 10$.

b. Note that $4 \mid (2 \cdot 6)$ but $4 \nmid 2$ and $4 \nmid 6$. This does not contradict Theorem 6.2.5 since neither $1 \neq (4,2)$ nor $1 \neq (4,6)$.

c. We showed in Example 6.2.3, that $4 = (48,68)$. Now $48 = a_1c = 12 \cdot 4$ and $68 = b_1c = 17 \cdot 4$. By Theorem 6.2.5, $1 = (12,17)$.

We shall now address the situation of finding all integers which have two given integers a and b as divisors (factors).

Definition 6.2.3

Let $a, b, m \in \mathbf{Z}$, where $a \neq 0$, $b \neq 0$, and $m > 0$. Then m is called a ***least common multiple (lcm)*** of a and b and we write $\mathbf{m = \langle a, b \rangle}$ if:

a. $a \mid m$ and $b \mid m$.

b. whenever $a \mid x$ and $b \mid x$, then $m \mid x$.

Note that if $m = \langle a, b \rangle$, then not only do a and b divide m, but m is the smallest positive integer into which both a and b will divide.

Example 6.2.6

a. Show $36 = \langle 12,18 \rangle$.

   i. $12 \mid 36$ since $36 = 12 \cdot 3$ and $18 \mid 36$ since $36 = 18 \cdot 2$.

   ii. Let $12 \mid x$ and $18 \mid x$. Since $12 \mid x$,
   $x \in \{0, \pm 12, \pm 24, \pm 36, \ldots\}$. Since $18 \mid x$,
   $x \in \{0, \pm 18, \pm 36, \pm 54, \ldots\}$. Hence, if $12 \mid x$ and $18 \mid x$, then
   $x \in \{0, \pm 12, \pm 24, \pm 36, \ldots\} \cap \{0, \pm 18, \pm 36, \pm 54, \ldots\} = \{0, \pm 36, \pm 72, \ldots\}$.

   Therefore $36 \mid x$.

b. Show $12 = \langle -4, 6 \rangle$

i. $(-4) \mid 12$ since $12 = (-4) \cdot (-3)$ and $6 \mid 12$ since $12 = 6 \cdot 2$.

ii. Let $(-4) \mid x$ and $6 \mid x$. Since $(-4) \mid x$ then, $x \in \{0, \pm4, \pm8, \pm12, \ldots\}$. Since $6 \mid x$, then $x \in \{0, \pm6, \pm12, \pm18, \ldots\}$. Hence, if $(-4) \mid x$ and $6 \mid x$, then $x \in \{0, \pm4, \pm8, \pm12, \ldots\} \cap \{0, \pm6, \pm12, \pm18, \ldots\} = \{0, \pm12, \pm24, \ldots\}$.

Therefore, $12 \mid x$.

The following theorem shows the relationship between the greatest common divisor and the least common multiple of two integers.

Theorem 6.2.6

Let $a, b \in \mathbf{Z}$ where $a \neq 0$ and $b \neq 0$. If $m = \frac{|ab|}{(a,b)}$, then $m = \langle a, b \rangle$.

Proof

Let $a, b \in \mathbf{Z}$, where $a \neq 0$ and $b \neq 0$. Then by Theorem 6.2.2, a and b have a gcd, (a,b).

Let $m = \frac{|ab|}{(a,b)}$. We wish to show $m = \langle a, b \rangle$. First note that $m > 0$.

1. **a | m and b | m**: Since $(a,b) \mid a$ and $(a,b) \mid b$, $(a,b) \mid (|a|)$ and $(a,b) \mid (|b|)$. Hence, $\exists u, v \in \mathbf{Z}$ such that $|a| = (a,b) \cdot u$ and $|b| = (a,b) \cdot v$ or $u = \frac{|a|}{(a,b)}$ and $v = \frac{|b|}{(a,b)}$. Now

$$m = \frac{|ab|}{(a,b)} = \frac{|a| \cdot |b|}{(a,b)} = |a| \cdot \frac{|b|}{(a,b)} = |a| \cdot v \text{ and}$$
$$m = \frac{|ab|}{(a,b)} = \frac{|a| \cdot |b|}{(a,b)} = |b| \cdot \frac{|a|}{(a,b)} = |b| \cdot u.$$

Thus we have $(|a|) \mid m$ and $(|b|) \mid m$ which implies that $a \mid m$ and $b \mid m$.

Proof continued next page.

2. **Whenever a | x and b | x, then m | x**: Let a | x and b | x. Then $(|a|) \mid x$ and $(|b|) \mid x$ and thus $\exists r, s \in \mathbf{Z}$ such that $x = |a| \bullet r$ and $x = |b| \bullet s$. If $c = (a,b)$, then $a = a_1c$ and $b = b_1c$ where $1 = (a_1,b_1)$. Since $|a| \bullet r = |b| \bullet s$, we have $|a_1c| \bullet r = |b_1c| \bullet s$ or $|a_1| \bullet r = |b_1| \bullet s$. Thus $(|a_1|) \mid (|b_1| \bullet s)$. But, if $1 = (a_1,b_1)$, then $1 = (|a_1|, |b_1|)$; and hence by Theorem 6.2.5, part (a), $(|a_1|) \mid s$. Therefore, $\exists d \in \mathbf{Z}$ such that $s = |a_1| \bullet d$ and we have

$$\begin{aligned} x &= |b| \bullet s \\ &= |b| \bullet (|a_1| \bullet d) && \text{since } s = |a_1| \bullet d \\ &= (|b| \bullet |a_1|) \bullet d \\ &= |ba_1| \bullet d \\ &= |b_1ca_1| \bullet d && \text{since } b = b_1c \\ &= |a_1b_1c| \bullet d \end{aligned}$$

Now $m = \dfrac{|ab|}{(a,b)} = \dfrac{|a_1cb_1c|}{c} = \dfrac{|a_1b_1c| \bullet |c|}{c} = \dfrac{|a_1b_1c| \bullet c}{c} = |a_1b_1c|$.

Thus $x = |a_1b_1c| \bullet d = md$ and we have $m \mid x$.

■

Example 6.2.7

Refer to Example 6.2.3

a. Find m = <48, 68>.

Since 4 = (48,68),

$$\begin{aligned} m &= \frac{|48 \bullet 68|}{(48,6)} \\ &= \frac{|3264|}{4} \\ &= \frac{3264}{4} \\ &= 816 \end{aligned}$$

b. Find $m = \langle -24, 63 \rangle$.

Since $3 = (-24,63)$,

$$m = \frac{|(-24) \cdot 63|}{(-24,63)} = \frac{|-1512|}{3} = \frac{1512}{3} = 504$$

Theorem 6.2.7

Let $a, b, m \in \mathbf{Z}$.

a. If $a \neq 0$ and $b \neq 0$, then a and b have a lcm, m.

b. If $m = \langle a, b \rangle$, then m is unique.

c. If $m = \langle a, b \rangle$, then $m = \langle |a|, |b| \rangle$. (In particular, if $m = \langle a, b \rangle$, then $m = \langle -a, b \rangle = \langle a, -b \rangle = \langle -a, -b \rangle$.)

Proof

See Exercises 6.2, problem (3).

■

Let $a, b \in \mathbf{Z}$ where $a \neq 0$ and $b \neq 0$. Then a and b have a lcm, m.
To find m, either:

find the smallest positive integer into which both a and b will divide; *or*
find the gcd of a and b and apply Theorem 6.2.6.

The concepts of gcd and lcm can be extended to include more than two integers. For example $6 = (12,24,30)$ while $120 = \langle 12, 24, 30 \rangle$.

Exercises 6.2

1. Prove parts (b) and (c) of Theorem 6.2.1.

2. Let $a, b \in \mathbf{Z}$ where $a \neq 0$ or $b \neq 0$. Prove that if $c = (a,b)$ and $z = ax + by$ for any $x, y \in \mathbf{Z}$, then $c \mid z$.

3. Prove Theorem 6.2.7

4. Let $a = 112$ and $b = 264$.

   a. Use the Euclidean Algorithm to find (112,264).

   b. Find $u, v \in \mathbf{Z}$ such that $(112,264) = 112u + 264v$.

   c. Find $<112, 264>$.

5. Let $a = 210$ and $b = -240$.

   a. Use the Euclidean Algorithm to find (210,-240).

   b. Find $u, v \in \mathbf{Z}$ such that $(210,-240) = 210u + (-240)v$.

   c. Find $<210, -240>$.

6. If $b \mid a$ and $c \mid a$ and $1 = (b, c)$, show $bc \mid a$.

7. Let a and b be two fixed integers such that $a, b \geq 2$. By Exercises 6.1, (6), the functions $f: \mathbf{N} \to \mathbf{N}$ where $f(n) = a^n$ and $g: \mathbf{N} \to \mathbf{N}$ where $g(n) = b^n$ are both 1–1. If $1 = (a,b)$, show that the function $h: \mathbf{N} \times \mathbf{N} \to \mathbf{N}$ where $h((m,n)) = a^m b^n \; \forall (m,n) \in \mathbf{N} \times \mathbf{N}$, is 1–1.

Section 6.3
PRIMES AND THE FUNDAMENTAL THEOREM OF ARITHMETIC

Prime integers have been referred to as the "building blocks" of the set of integers. If an integer n, $n \neq 0$, $n \neq \pm 1$, is not prime, then we shall see that it can be factored into a product of primes in basically only one way.

Definition 6.3.1

Let $n \in \mathbf{Z}$, where $n \neq 0$, $n \neq \pm 1$.

a. n is called a ***prime*** integer if the only divisors of n are ± 1 and $\pm n$.

b. n is called a ***composite*** integer if n is not a prime integer; i.e., if n has divisors other than ± 1 and $\pm n$.

Example 6.3.1

a. $n = -5$ is a prime integer since $-5 \neq 0, \pm 1$ and the only divisors of -5 are ± 1 and ± 5.

b. $n = 12$ is a composite integer since $12 \neq 0, \pm 1$, and 12 has divisors other than ± 1 and ± 12; i.e., $\pm 2, \pm 3, \pm 4, \pm 6$.

Discussion

If one wishes to show an integer n is prime, one must show:

$n \neq 0$, $n \neq \pm 1$; *and*

if $k \mid n$, $k \in \mathbf{Z}$, then $k = \pm 1$ or $\pm n$.

Theorem 6.3.1

a. If p is a prime integer, then -p is a prime integer.

b. The only even prime integers are ± 2.

c. If $n \in \mathbf{Z}$, where $n \neq 0, \pm 1$, then there exists a prime integer p such that $p \mid n$.

Proof

We shall prove parts (b) and (c) and leave the proof of part (a) as an exercise.

b. By part (a) it suffices to show that 2 is the only even, positive prime integer. Let p be an even, positive prime integer. Since p is even, $p = 2n$ where $n \in \mathbf{Z}^+$ and $1 \leq n < p$. Thus $n \mid p$; and, since p is prime, $n = 1$ or $n = p$. But $n < p$. Therefore, $n = 1$ and $p = 2 \cdot 1 = 2$.

■

c. By Theorem 6.1.3, parts (a) and (b), it suffices to show that if $n \in \mathbf{Z}$ and $n \geq 2$, then there exists a positive prime integer p such that $p \mid n$. The desired result follows from Example 1.5.3, part (b).

■

The following result provides insight into determining whether an integer is prime or composite.

Theorem 6.3.2

Let $n \in \mathbf{Z}$ where $n > 1$. If n is a composite integer, then there exists a positive prime integer p such that $p \mid n$ and $p \leq \sqrt{n}$.

Proof

Let $n \in \mathbf{Z}$ where $n > 1$ and n is a composite. Then $\exists a, b \in \mathbf{Z}^+$ such that $n = ab$ and $1 < a, b < n$. If both $a > \sqrt{n}$ and $b > \sqrt{n}$, then $ab > \sqrt{n} \cdot \sqrt{n}$ or $ab > n$. Hence, either $a \leq \sqrt{n}$ or $b \leq \sqrt{n}$. We may assume without loss of generality that $a \leq \sqrt{n}$. By Theorem 6.3.1, since $a > 1$, there exists a prime integer and hence a positive prime integer, p, such that $p \mid a$. Since $p \mid a$ and $a \mid n$, $p \mid n$. Also, since $p \mid a$ and $a > 1$, by Theorem 6.1.3, part (c), $p \leq |a| = a$. Therefore, $p \leq a$ and $a \leq \sqrt{n}$, which implies $p \leq \sqrt{n}$.

■

Example 6.3.2

a. Let $n = 12$. Then 12 is a composite integer such that $2 \mid 2$ and $3 \mid 12$. Now $2 \leq \sqrt{12}$ and $3 \leq \sqrt{12}$ and hence we have two choices for p; i.e., $p = 2$ or $p = 3$.

b. Let $n = 15$. Then 15 is a composite integer such that $3 \mid 15$ and $5 \mid 15$. Now $3 \leq \sqrt{15}$ and $5 \nleq \sqrt{15}$. Hence, we have only one choice for p; i.e., $p = 3$.

c. Let $n = 103$. We wish to determine if 103 is a prime or composite integer. Assume 103 is a composite integer. Then by Theorem 6.3.2 there must exist a positive prime integer p such that $p \leq \sqrt{103}$ (or $p^2 \leq 103$) and $p \mid 103$. Now $2^2 = 4$, $3^2 = 9$, $5^2 = 25$, $7^2 = 49$ and $11^2 = 121$. Hence, 7 is the largest prime integer such that $7^2 < 103$. Thus p must be 2, 3, 5, or 7. But $2 \nmid 103$, $3 \nmid 103$, $5 \nmid 103$ and $7 \nmid 103$ which contradicts Theorem 6.3.2. Hence, we conclude 103 is a prime integer.

Theorem 6.3.3

a. Let $a, p \in \mathbf{Z}$. If p is a prime integer such that $p \nmid a$, then $1 = (p,a)$.

b. Let $a, b, p \in \mathbf{Z}$. If p is a prime integer and $p \mid ab$, then $p \mid a$ or $p \mid b$.

c. Let $a_1, \ldots, a_n, p \in \mathbf{Z}$ where $n \geq 2$. If p is a prime integer and $p \mid a_1 a_2 \cdots a_n$, then $p \mid a_i$ for some i, $1 \leq i \leq n$.

Proof

We shall prove parts (a) and (b) and leave the proof of part (c) as an exercise.

a. Let $a, p \in \mathbf{Z}$ where p is a prime integer and $p \nmid a$. Now $p \neq 0$; and, since $p \nmid a$, $a \neq 0$. Hence, the pair of integers a and p have a gcd, say c. Now $c \mid p$ and thus $c = 1$ or $c = \pm p$. Since $c = (a,p)$ and $p \nmid a$, $c \neq \pm p$. Therefore, $c = 1$ and we have $1 = (p,a)$.

■

b. Let $a, b, p \in \mathbf{Z}$ where p is a prime integer and $p \mid ab$. Suppose $p \nmid a$. Then $1 = (p,a)$. By Theorem 6.2.5, part (a), $p \mid b$.

■

We may have $a \mid bc$ even though $a \nmid b$ and $a \nmid c$. For example $6 \mid (3 \cdot 4)$ but $6 \nmid 3$ and $6 \nmid 4$. This does not contradict Theorem 6.3.3, part (b), since 6 is not a prime integer.

Theorem 6.3.4

a. Let $p, q \in \mathbf{Z}$ where p and q are prime integers. If $p \mid q$, then $p = \pm q$.

b. Let $q_1, \ldots, q_n, p \in \mathbf{Z}$ where $q_1, \ldots, q_n$ and p are prime integers. If $p \mid q_1 \cdots q_n$, then $p = \pm q_i$ for some i, $1 \leq i \leq n$.

Proof

a. Let $p, q \in \mathbf{Z}$ where p and q are prime integers. Suppose $p \mid q$. Since q is a prime integer, $p = \pm 1$ or $p = \pm q$. Since p is a prime integer, $p \neq \pm 1$. Therefore $p = \pm q$.

■

b. Let $q_1, \ldots, q_n, p \in \mathbf{Z}$ where $q_1, \ldots, q_n$ and p are prime integers. Suppose $p \mid q_1 q_2 \cdots q_n$. By Theorem 6.3.3, part (c), $p \mid q_i$ for some i, $1 \leq i \leq n$. Hence, by part (a) above, $p = \pm q_i$.

■

The following result is also referred to sometimes as the *Unique Factorization Theorem*. It provides the blueprint for how the primes can be used as "building blocks" for the integers.

Theorem 6.3.5 (Fundamental Theorem of Arithmetic)

Let $n \in \mathbf{Z}$ where $n \neq 0$, $n \neq \pm 1$.

a. Then n is either a prime integer or n can be expressed in the form $n = p_1 p_2 \cdots p_r$ if $n > 1$ or $n = -(p_1 p_2 \cdots p_r)$ if $n < -1$ where p_i, $1 \leq i \leq r$, are positive prime integers.

b. The representation of n as a product of positive prime integers is unique except for order. That is, if $n = \pm(p_1 \cdots p_r)$ and $n = \pm(q_1 \cdots q_s)$ where the p_i and q_j are positive prime integers, then $r = s$ and for each i, $1 \leq i \leq r$, $p_i = q_j$.

Proof

It suffices to prove the theorem is true for $n \in \mathbf{Z}$, $n \geq 2$.

a. We shall use Mathematical Induction II. For $n \geq 2$, let p(n) be the statement that n can be factored into a product of a finite number of positive prime integers.

p(2): Since 2 is prime and $2 \mid 2$, 2 has a factorization with one prime factor. Therefore, p(2) is true.

p(k): Assume p(k) is true for all $k \in \mathbf{Z}$, $2 \leq k < m$. Then each k, $2 \leq k < m$, can be factored into a finite product of positive prime integers.

p(m): Show p(m) is true; i.e., m can be factored into a finite product of positive prime integers.

If m is prime, them $m \mid m$, and m has a factorization with one prime factor. If m is not prime, them $m = st$ where $1 < s, t < m$. Since $2 < s, t < m$, p(s) and p(t) are true; i.e., s and t both can be factored into a finite product of positive prime integers.

Hence, $m = st$
$= (p_1 \cdots p_u)(q_1 \cdots q_v)$

where the p_i and q_j are positive prime integers. In either case, p(m) is true.

Hence, $(\forall n, n \geq 2)$ p(n) is true.

■

b. Suppose $n = p_1 \cdots p_r$, $n = q_1 \cdots q_s$ where the p_i and q_j are positive prime integers and $r \neq s$. Without loss of generality, we may assume $r < s$. Now, since $p_1p_2 \cdots p_r = q_1q_2 \cdots q_s$, $p_1 \mid q_1q_2 \cdots q_s$. By Theorem 6.3.4, part (b), $p_1 = q_j$ for some j, say $p_1 = q_1$. Therefore we have $p_1p_2 \cdots p_r = p_1q_2 \cdots q_s$ or $p_2p_3 \cdots p_r = q_2q_3 \cdots q_s$. Again, $p_2 \mid q_2q_3 \cdots q_s$ and hence $p_2 = q_j$ for some j, say $p_2 = q_2$. We thus have $p_2p_3 \cdots p_r = p_2q_3 \cdots q_s$ or $p_3 \cdots p_r = q_3 \cdots q_s$. By repeated application of Theorem 6.3.4, part (b), we get $p_1 = q_1, p_2 = q_2, \ldots, p_r = q_r$; and, since $r < s$, $1 = q_{r+1} \cdots q_s$. But $1 = q_{r+1} \cdots q_s$ and $q_j > 0$ imply $q_{r+1} = \cdots = q_s = 1$. Hence, $q_{r+1}, \cdots, q_s$ are not prime. This is a contradiction which was arrived at by assuming $r \neq s$. Thus $r = s$.

■

Let $n \in \mathbf{Z}$, where $n = \pm(p_1 \cdots p_r)$. Since $p_1, p_2, \cdots, p_r$ are not necessarily distinct, we may rewrite the factorization of n as follows:

$$n = \pm(p_1^{e_1} p_2^{e_2} \cdots p_k^{e_k})$$

where $p_1, p_2, \ldots, p_k$ are distinct primes and $e_1, e_2, \cdots, e_k \in \mathbf{Z}^+$.

Example 6.3.3

Express n as a product of positive prime integers.

a. $n = 48$
$$\begin{aligned} &= 2 \cdot 24 \\ &= 2 \cdot 2 \cdot 12 \\ &= 2 \cdot 2 \cdot 2 \cdot 6 \\ &= 2 \cdot 2 \cdot 2 \cdot 2 \cdot 3 \\ &= 2^4 \cdot 3^1. \end{aligned}$$

a. $n = -612$
$$\begin{aligned} &= -(612) \\ &= -(2 \cdot 306) \\ &= -(2 \cdot 2 \cdot 153) \\ &= -(2 \cdot 2 \cdot 3 \cdot 51) \\ &= -(2 \cdot 2 \cdot 3 \cdot 3 \cdot 17) \\ &= -(2^2 \cdot 3^2 \cdot 17^1). \end{aligned}$$

Let $a, b \in \mathbf{Z}$ where $a, b \neq 0, \pm 1$. Then by the Fundamental Theorem of Arithmetic,

$$a = \pm(p_1^{e_1} p_2^{e_2} \bullet\bullet\bullet\ p_k^{e_k}) \text{ and } b = \pm(q_1^{f_1} q_2^{f_2} \bullet\bullet\bullet\ q_r^{f_r}).$$

Let $S = \{p_1, \bullet\bullet\bullet, p_k\} \cap \{q_1, \bullet\bullet\bullet, p_r\}$ and $T = \{p_1, \bullet\bullet\bullet, p_k\} \cup \{q_1, \bullet\bullet\bullet, p_r\}$.

If $c = (a,b)$ and $c \neq 1$, then c can be expressed in the form:

$$c = p_1^{s_1} \bullet\bullet\bullet\ p_t^{s_t} \text{ where } p_i \in S \text{ and } s_i = \min\{e_i, f_i\}.$$

If $m = <a,b>$, then m can be expressed in the form:

$$m = p_1^{t_1} p_2^{t_2} \bullet\bullet\bullet\ p_v^{t_v} \text{ where } p_i \in T \text{ and } t_i = \max\{e_i, f_i\}.$$

[Note: If p_i is a prime factor of a but not b (or vice versa),
then take $f_i = 0$ $(e_i = 0)$.]

Example 6.3.4

Let $a = 48$ and $b = -612$.

Now $48 = 2^4 \bullet 3^1$ and $-612 = -(2^2 \bullet 3^2 \bullet 17^1)$. Hence,
$S = \{2,3\} \cap \{2,3,17\} = \{2,3\}$ and $T = \{2,3\} \cup \{2,3,17\} = \{2,3,17\}$.

$$\begin{aligned}\text{Thus } c = (48,-612) &= 2^{s_1} \bullet 3^{s_2} \\ &= 2^{\min\{4,2\}} \bullet 3^{\min\{1,2\}} \\ &= 2^2 \bullet 3^1 \\ &= 4 \bullet 3 \\ &= 12\end{aligned}$$

$$\begin{aligned}m = <48,-612> &= 2^{t_1} \bullet 3^{t_2} \bullet 17^{t_3} \\ &= 2^{\max\{4,2\}} \bullet 3^{\max\{1,2\}}\ 17^{\max\{0,1\}} \\ &= 2^4 \bullet 3^2 \bullet 17^1 \\ &= 16 \bullet 9 \bullet 17 \\ &= 2448\end{aligned}$$

Exercises 6.3

1. Prove part (a) of Theorem 6.3.1.

2. Prove part (c) of Theorem 6.3.3. (Hint: Use Mathematical Induction I where $n \geq 2$.)

3. Use Theorem 6.3.2 to determine if n is a prime or composite integer.
   a. $n = 201$
   b. $n = 293$

4. Apply the Fundamental Theorem of Arithmetic to the following integers n.
   a. $n = 19500$
   b. $n = -252$

5. Let $a = 112$ and $b = 264$.
   a. Find the prime factorizations of a and b.
   b. Use the results of part (a) to find (112,264) and <112,264>.
   c. Compare your results of part (b) with Exercises 6.2, problem (4).

6. Let $a = 210$ and $b = -240$.
   a. Find the prime factorizations of a and b.
   b. Use the results of part (a) to find (210,-240) and <210,-240>.
   c. Compare your results of part (b) with Exercises 6.2, problem (5).

Section 6.4
CONGRUENCES

In Chapter 4, Section 4.3, we saw that an equivalence relation, $\mathcal{R}$, defined on a set A generates a family of subsets of A (the equivalence sets determined by $\mathcal{R}$) which forms a partition of A. Furthermore, elements of A that are related under $\mathcal{R}$ belong to the same equivalence set and hence can be considered as being equivalent.

We now wish to turn our attention to a special equivalence relation defined on **Z**. In Example 4.3.2, part (c), we defined the relation $\mathcal{R}$ on **Z** as follows:

$$\text{For } a, b \in \mathbf{Z},\ a\,\mathcal{R}\,b \Leftrightarrow a - b = 3k,\ k \in \mathbf{Z} \Leftrightarrow 3 \mid (a - b).$$

We showed $\mathcal{R}$ is an equivalence relation. This is a particular example of an equivalence relation called a *congruence*.

Definition 6.4.1

Let $m \in \mathbf{Z}$, $m > 1$. For $a, b \in \mathbf{Z}$, we shall say a is ***congruent*** to b modulo m and we write $\mathbf{a \equiv b \ (mod\ m)}$ if $m \mid (a - b)$.

Example 6.4.1

a. Let $m = 5$.

$12 \equiv 2 \pmod 5$ since $5 \mid (12 - 2)$ or $5 \mid 10$.

$-8 \equiv 7 \pmod 5$ since $5 \mid (-8 - 7)$ or $5 \mid (-15)$.

$2 \not\equiv (-4) \pmod 5$ since $5 \nmid [2 - (-4)]$ or $5 \nmid 6$.

b. Let $m = 6$.

$27 \equiv (-3) \pmod 6$ since $6 \mid [27 - (-3)]$ or $6 \mid 30$.

$4 \not\equiv (-3) \pmod 6$ since $6 \nmid [4 - (-3)]$ or $6 \nmid 7$.

As we proceed through this section, it will become evident that the properties possessed by congruences are quite similar in nature to those of the equality relation on the integers. The first of these properties is exhibited by the following theorem.

Theorem 6.4.1

Let $m \in \mathbf{Z}$, $m > 1$. For $a, b, c, d \in \mathbf{Z}$, if $a \equiv c \pmod m$ and $b \equiv d \pmod m$, then

a. $(a + b) \equiv (c + d) \pmod m$.

b. $ab \equiv cd \pmod m$.

Proof

We shall prove part (b) and leave the proof of part (a) as an exercise.

b. Let $a, b, c, d \in \mathbf{Z}$ such that $a \equiv c \pmod{m}$ and $b \equiv d \pmod{m}$. Then $m \mid (a - c)$ and $m \mid (b - d)$ which implies $\exists r, s \in \mathbf{Z}$ such that $a - c = rm$ and $b - d = sm$ or $a = c + rm$ and $b = d + sm$. Hence,

$$\begin{aligned} ab &= (c + rm)(d + sm) \\ &= cd + c(sm) + (rm)d + (rm)(sm) \\ &= cd + (cs)m + (rd)m + (rms)m \\ &= cd + (cs + rd + rms)m \\ &= cd + tm \quad \text{where } t = cs + rd + rms \in \mathbf{Z}. \end{aligned}$$

Therefore, $ab - cd = tm$ or $m \mid (ab - cd)$. Thus, $ab \equiv cd \pmod{m}$.

■

Theorem 6.4.2

Let $m \in \mathbf{Z}$, $m > 1$. If $a, b \in \mathbf{Z}$ and $1 = (a,m)$, then there exists $x \in \mathbf{Z}$ such that $ax \equiv b \pmod{m}$.

Proof

Let $a, b \in \mathbf{Z}$ such that $1 = (a,m)$.

Now $\exists u, v \in \mathbf{Z}$ such that $1 = au + mv$.

Hence,

$$\begin{aligned} 1 &= au + mv \\ b &= b(au + mv) \\ &= a(bu) + (bv)m \end{aligned}$$

or

$$\begin{aligned} a(bu) &= b - (bv)m \\ &= b + [-(bv)]m \\ &= b + tm \qquad \text{where } t = -(bv) \in \mathbf{Z}. \end{aligned}$$

Let $x = bu$. Then $x \in \mathbf{Z}$ and $ax = b + tm$ or $ax - b = tm$. Therefore, $m \mid (ax - b)$ and $ax \equiv b \pmod{m}$.

■

Example 6.4.2

Let $m = 5$.

a. From Example 6.4.1, we have $12 \equiv 2 \pmod{5}$ and $-8 \equiv 7 \pmod{5}$. Hence, by Theorem 6.4.1, we have

$$[12 + (-8)] \equiv [2 + 7] \pmod{5} \text{ or } 4 \equiv 9 \pmod{5}.$$

and

$$12 \cdot (-8) \equiv 2 \cdot 7 \pmod{5} \text{ or } -96 \equiv 14 \pmod{5}.$$

b. Let $a = 2$ and $b = 3$. Since $1 = (2,5)$, by Theorem 6.4.2, $\exists x \in \mathbf{Z}$ such that $2x \equiv 3 \pmod{5}$. If $2x \equiv 3 \pmod{5}$, then $5 \mid (2x - 3)$. Let $x = 4$. Then $5 \mid (8 - 3)$ or $5 \mid 5$. In fact, if $x \in \{\ldots, -6, -1, 4, 9, 14, \ldots\}$, then the relation $2x \equiv 3 \pmod{5}$ will hold.

Theorem 6.4.3

Let $m \in \mathbf{Z}$, where $m > 1$. Then congruence mod m, denoted by $\equiv$, is an equivalence relation on $\mathbf{Z}$.

Proof

Let $m \in \mathbf{Z}$, where $m > 1$.

a. Let $a \in \mathbf{Z}$. Since $m \neq 0$, $m \mid 0$. Thus $m \mid (a - a)$ which implies $a \equiv a \pmod{m}$. Therefore, $\equiv$ is reflexive.

b. Let $a \equiv b \pmod{m}$. Then we have $m \mid (a - b)$. Thus $m \mid [-(a - b)]$ or $m \mid (b - a)$ which implies $b \equiv a \pmod{m}$. Therefore, $\equiv$ is symmetric.

c. Let $a \equiv b \pmod{m}$ and $b \equiv c \pmod{m}$. Then $m \mid (a - b)$ and $m \mid b - c$. By Theorem 6.1.3, part (g), $m \mid [(a - b) + (b - c)]$ or $m \mid (a - c)$. Thus $a \equiv c \pmod{m}$. Therefore, $\equiv$ is transitive.

Hence, congruence mod m, $\equiv$, is an equivalence relation on $\mathbf{Z}$.

■

Let $m \in \mathbf{Z}$, $m > 1$. Since congruence mod m is an equivalence relation on $\mathbf{Z}$, $\forall a \in \mathbf{Z}$, we have the equivalence sets $\overline{a} = \{x \in \mathbf{Z} \mid x \equiv a \pmod{m}\}$. Now

$\{\bar{a} \mid a \in \mathbf{Z}\}$ forms a partition of $\mathbf{Z}$ (see Theorem 4.5.1). Hence, $\mathbf{Z} = \bigcup_{a \in \mathbf{Z}} \bar{a}$; and, for $\bar{a}, \bar{b} \in \{\bar{a} \mid a \in \mathbf{Z}\}$, either $\bar{a} = \bar{b}$ or $\bar{a} \cap \bar{b} = \emptyset$. Let $\mathbf{Z}_m = \{\bar{a} \mid a \in \mathbf{Z}\}$. $\mathbf{Z}_m$ is called the *set of integers modulo m.*

Example 6.4.3

Let m = 5. Then $\mathbf{Z}_5 = \{\bar{a} \mid a \in \mathbf{Z}\} = \{\ldots, \overline{-2}, \overline{-1}, \bar{0}, \bar{1}, \bar{2}, \ldots\}$ is the set of integers modulo 5. Now

$$\begin{aligned}\overline{-2} &= \{x \in \mathbf{Z} \mid x \equiv (-2) \pmod 5\}\\ &= \{x \in \mathbf{Z} \mid 5 \mid (x + 2)\}\\ &= \{\ldots, -12, -7, -2, 3, 8, \ldots\}\end{aligned}$$

$$\begin{aligned}\bar{0} &= \{x \in \mathbf{Z} \mid x \equiv 0 \pmod 5\}\\ &= \{x \in \mathbf{Z} \mid 5 \mid x\}\\ &= \{\ldots, -10, -5, 0, 5, 10, \ldots\}\end{aligned}$$

$$\begin{aligned}\bar{1} &= \{x \in \mathbf{Z} \mid x \equiv 1 \pmod 5\}\\ &= \{x \in \mathbf{Z} \mid 5 \mid (x - 1)\}\\ &= \{\ldots, -9, -4, 1, 6, 11, \ldots\}\end{aligned}$$

Theorem 6.4.4

If $m \in \mathbf{Z}$, $m > 1$, then there exists exactly m distinct elements of $\mathbf{Z}_m$; i.e., $\mathbf{Z}_m = \{\bar{0}, \bar{1}, \ldots, \overline{(m-1)}\}$.

Proof

Let $m \in \mathbf{Z}$ where $m > 1$. We must first show that the m elements $\bar{0}, \bar{1}, \ldots, \overline{(m-1)}$ are all distinct and then show $\mathbf{Z}_m = \{\bar{0}, \bar{1}, \ldots, \overline{(m-1)}\}$.

1. Let $a, b \in \mathbf{Z}$ where $0 \le a, b < m$ and $a \ne b$. Since $a \ne b$, either $a < b$ or $a > b$. We may assume without loss of generality that $a > b$. Then $0 < a - b < m$. Therefore, $m \nmid (a - b)$ and $a \not\equiv b \pmod m$. Hence, by Theorem 4.4.1, part (d), $\bar{a} \ne \bar{b}$. Thus the m elements $\bar{0}, \bar{1}, \ldots, \overline{(m-1)}$ are all distinct.

Proof continued next page.

2. Since $\{\overline{0}, \overline{1}, \ldots, \overline{(m-1)}\} \subseteq \mathbf{Z}_m$, we must show that $\mathbf{Z}_m \subseteq \{\overline{0}, \overline{1}, \ldots, \overline{(m-1)}\}$. Let $\overline{a} \in \mathbf{Z}_m$. Then $a \in \mathbf{Z}$; and, by the division algorithm, we have $a = mq + r$ where $0 \leq r < |m| = m$. Therefore, $a - r = mq$ and $m \mid (a - r)$ which implies $a \equiv r \pmod{m}$ yielding $\overline{a} = \overline{r}$ by Theorem 4.4.1, part (d). Since $0 \leq r < m, \overline{r} \in \{\overline{0}, \overline{1}, \ldots, \overline{(m-1)}\}$. Thus $\overline{a} \in \{\overline{0}, \overline{1}, \ldots, \overline{(m-1)}\}$. Hence $\mathbf{Z}_m = \{\overline{0}, \overline{1}, \ldots, \overline{(m-1)}\}$.

■

Example 6.4.4

For m = 4, we have $\mathbf{Z}_4 = \{\overline{0}, \overline{1}, \overline{2}, \overline{3}\}$ where

$\overline{0} = \{\ldots, -8, -4, 0, 4, 8, \ldots\}$

$\overline{1} = \{\ldots, -7, -3, 1, 5, 9, \ldots\}$

$\overline{2} = \{\ldots, -6, -2, 2, 6, 10, \ldots\}$

$\overline{3} = \{\ldots, -5, -1, 3, 7, 11, \ldots\}$

The proof of Theorem 6.4.4, part (b), shows how an equivalence set $\overline{a}$ can be equated to an equivalence set $\overline{r}$ where $0 \leq r < m$. If $a = -5$, then by the division algorithm $-5 = (-2) \cdot 4 + 3$ and $0 < 3 < 4$. Hence $\overline{-5} = \overline{3}$.

It is useful to note that two integers a and b are congruent modulo m if and only if they leave the same nonnegative remainder r when divided by m. Consequently, by Theorem 6.4.4, $\overline{a} = \overline{r} = \overline{b}$. For $m = 4$ in Example 6.4.4, we note $7 = 1 \cdot 4 + 3$ and $11 = 2 \cdot 4 + 3$. Thus $\overline{7} = \overline{3} = \overline{11}$. For this reason, the equivalence sets determined by congruence modulo m are often called *residue classes.*

Let $m \in \mathbf{Z}$, $m > 1$. Note that if $\overline{a}, \overline{b} \in \mathbf{Z}_m$, then $a, b \in \mathbf{Z}$ and hence $a + b, a \cdot b \in \mathbf{Z}$. Thus we may consider the equivalence sets $\overline{a+b}$ and $\overline{a \cdot b}$ which are again elements of $\mathbf{Z}_m$. This observation leads to the following definitions: For $\overline{a}, \overline{b} \in \mathbf{Z}_m$, let

$$\overline{a} + \overline{b} = \overline{a+b} \quad \text{and} \quad \overline{a} \cdot \overline{b} = \overline{a \cdot b}.$$

The question is whether or not the + and • as defined on $\mathbf{Z}_m$ are operations on $\mathbf{Z}_m$. That is, are + and • functions from $\mathbf{Z}_m \times \mathbf{Z}_m \to \mathbf{Z}_m$? The difficulty arises from the fact that we are dealing with equivalence sets and hence an element of

$\mathbf{Z}_m \times \mathbf{Z}_m$ may have more than one representation (see Theorem 4.4.1, part (d)). For example, in $\mathbf{Z}_4 \times \mathbf{Z}_4$ we have $(\overline{1}, \overline{3}) = (\overline{5}, \overline{7})$ because $\overline{1} = \overline{5}$ ($1 \equiv 5 \pmod 4$) and $\overline{3} = \overline{7}$ ($3 \equiv 7 \pmod 4$). We must make sure that regardless of which representation is used for a given element of $\mathbf{Z}_m \times \mathbf{Z}_m$, the image of that element under + or • is unique; that is, + and • are well defined.

Theorem 6.4.5

Let $m \in \mathbf{Z}$, $m > 1$. If for $\overline{a}, \overline{b} \in \mathbf{Z}_m$, $\overline{a} + \overline{b} = \overline{a + b}$ and $\overline{a} \bullet \overline{b} = \overline{a \bullet b}$, then + and • are operations on $\mathbf{Z}_m$.

Proof

Let $(\overline{a}, \overline{b}), (\overline{c}, \overline{d}) \in \mathbf{Z}_m \times \mathbf{Z}_m$ such that $(\overline{a}, \overline{b}) = (\overline{c}, \overline{d})$. Then $\overline{a} = \overline{c}$ and $\overline{b} = \overline{d}$; and; by Theorem 4.4.1, part (d), $a \equiv c \pmod m$ and $b \equiv d \pmod m$.

To show + is an operation on $\mathbf{Z}_m$ we must show that $\overline{a} + \overline{b} = \overline{c} + \overline{d}$. By Theorem 6.4.1, part (a), $(a + b) \equiv (c + d) \pmod m$. Hence, $\overline{a + b} = \overline{c + d}$ and we have

$$\begin{aligned} \overline{a} + \overline{b} &= \overline{a + b} \\ &= \overline{c + d} \\ &= \overline{c} + \overline{d} \end{aligned}$$

To show • is an operation on $\mathbf{Z}_m$ we must show that $\overline{a} \bullet \overline{b} = \overline{c} \bullet \overline{d}$. By Theorem 6.4.1, part (b), $a \bullet b \equiv c \bullet d \pmod m$. Hence, $\overline{a \bullet b} = \overline{c \bullet d}$ and we have

$$\begin{aligned} \overline{a} \bullet \overline{b} &= \overline{a \bullet b} \\ &= \overline{c \bullet d} \\ &= \overline{c} \bullet \overline{d} \end{aligned}$$

■

Example 6.4.5

The following are table representations for the operations + and • on $\mathbf{Z}_3$ and $\mathbf{Z}_4$. (Recall that if $\overline{a}, \overline{b} \in \mathbf{Z}_m$, then $a, b \in \mathbf{Z}$ and $a + b, a \bullet b \in \mathbf{Z}$. By the division algorithm $a + b = mq_1 + r_1$ and $a \bullet b = mq_2 + r_2$ where $0 \le r_1, r_2 < |m| = m$. Hence, $\overline{a} + \overline{b} = \overline{a + b} = \overline{r_1}$ and $\overline{a} \bullet \overline{b} = \overline{a \bullet b} = \overline{r_2}$. Specifically, in $\mathbf{Z}_4$, $\overline{3} + \overline{3} = \overline{6} = \overline{2}$ since $6 = 4(1) + 2$ and and $\overline{2} \bullet \overline{2} = \overline{4} = \overline{0}$ since $4 = 4(1) + 0$.)

a. $\mathbf{Z}_3 = \{\overline{0}, \overline{1}, \overline{2}\}$

+	$\overline{0}$	$\overline{1}$	$\overline{2}$
$\overline{0}$	$\overline{0}$	$\overline{1}$	$\overline{2}$
$\overline{1}$	$\overline{1}$	2	$\overline{0}$
$\overline{2}$	$\overline{2}$	$\overline{0}$	$\overline{1}$

•	$\overline{0}$	$\overline{1}$	$\overline{2}$
$\overline{0}$	$\overline{0}$	$\overline{0}$	$\overline{0}$
$\overline{1}$	$\overline{0}$	$\overline{1}$	$\overline{2}$
$\overline{2}$	$\overline{0}$	$\overline{2}$	$\overline{1}$

b. $\mathbf{Z}_4 = \{\overline{0}, \overline{1}, \overline{2}, \overline{3}\}$

+	$\overline{0}$	$\overline{1}$	$\overline{2}$	$\overline{3}$
$\overline{0}$	$\overline{0}$	$\overline{1}$	$\overline{2}$	$\overline{3}$
$\overline{1}$	$\overline{1}$	$\overline{2}$	$\overline{3}$	$\overline{0}$
$\overline{2}$	$\overline{2}$	$\overline{3}$	$\overline{0}$	$\overline{1}$
$\overline{3}$	$\overline{3}$	$\overline{0}$	$\overline{1}$	$\overline{2}$

•	$\overline{0}$	$\overline{1}$	$\overline{2}$	$\overline{3}$
$\overline{0}$	$\overline{0}$	$\overline{0}$	$\overline{0}$	$\overline{0}$
$\overline{1}$	$\overline{0}$	$\overline{1}$	$\overline{2}$	$\overline{3}$
$\overline{2}$	$\overline{0}$	$\overline{2}$	$\overline{0}$	$\overline{2}$
$\overline{3}$	$\overline{0}$	$\overline{3}$	$\overline{2}$	$\overline{1}$

The next theorem shows that most of the arithmetic properties of the mathematical system $\mathbf{Z}_m$ are analogous to those of $\mathbf{Z}$.

Theorem 6.4.6

Let $m \in \mathbf{Z}$, $m > 1$ and consider the set $\mathbf{Z}_m$ with the operations + and •. Then,

a. + is commutative.

b. + is associative.

c. $\overline{0}$ is the identity element of $\mathbf{Z}_m$ relative to +.

d. Every element of $\mathbf{Z}_m$ has an inverse relative to +.

e. • is commutative.

f. • is associative.

g. $\overline{1}$ is the identity element of $\mathbf{Z}_m$ relative to •.

h. • is distributive over +.

Proof

We shall prove parts (a), (d), (f), and (g) and leave the proofs of parts (b), (c), (e) and (h) as exercises. The proofs rely on the properties of **Z** relative to its operations of + and • .

a. Let $\overline{a}, \overline{b} \in \mathbf{Z}_m$. Then
$$\begin{aligned} \overline{a} + \overline{b} &= \overline{a + b} \\ &= \overline{b + a} \\ &= \overline{b} + \overline{a} \end{aligned}$$
Hence, + is commutative. ■

d. Let $\overline{a} \in \mathbf{Z}_m$. Then $a \in \mathbf{Z}$ and hence $-a \in \mathbf{Z}$.
By the division algorithm $-a = mq + r$ where $0 \leq r < |m| = m$.
Therefore, $\overline{-a} = \overline{r}$. Now
$$\begin{aligned} \overline{a} + \overline{r} &= \overline{a} + \overline{-a} \\ &= \overline{a + (-a)} \\ &= \overline{0} \end{aligned}$$
Since + is commutative, $\overline{r} + \overline{a} = \overline{0}$. Hence, $\overline{a}$ has an inverse relative to +. ■

f. Let $\overline{a}, \overline{b}, \overline{c} \in \mathbf{Z}_m$. Then
$$\begin{aligned} \overline{a} \bullet (\overline{b} \bullet \overline{c}) &= \overline{a} \bullet (\overline{b \bullet c}) \\ &= \overline{a \bullet (b \bullet c)} \\ &= \overline{(a \bullet b) \bullet c} \\ &= (\overline{a \bullet b}) \bullet \overline{c} \\ &= (\overline{a} \bullet \overline{b}) \bullet \overline{c} \end{aligned}$$
Hence, • is associative. ■

g. Let $\overline{a} \in \mathbf{Z}_m$. Then
$$\begin{aligned} \overline{a} \bullet \overline{1} &= \overline{a \bullet 1} \\ &= \overline{a} \end{aligned}$$
Since • is commutative, $\overline{1} \bullet \overline{a} = \overline{a}$.
Hence, $\overline{1}$ is the identity element of $\mathbf{Z}_m$ relative to • ■

$\mathbf{Z}_m$ together with the operations of + and • is an example of a mathematical system called a *commutative ring with identity*.

A couple of observations regarding $\mathbf{Z}_m$ together with the operation • are in order. First note that for $\overline{a}, \overline{b} \in \mathbf{Z}_m$, it could happen that $\overline{a} \bullet \overline{b} = \overline{0}$ even though $\overline{a} \neq \overline{0}$ and $b \neq \overline{0}$. For example in $\mathbf{Z}_4$ we have $\overline{2} \bullet \overline{2} = \overline{0}$ and in $\mathbf{Z}_6$ we have $\overline{2} \bullet \overline{3} = \overline{0}$.

Secondly, note that while $\overline{1}$ is the identity element of $\mathbf{Z}_m$ relative to •, not every $\overline{a} \in \mathbf{Z}_m$ has an inverse relative to • . For example, in $\mathbf{Z}_4$ the element $\overline{2}$ does not have an inverse relative to • since there does not exist $\overline{x} \in \mathbf{Z}_4$ such that $\overline{2} \bullet \overline{x} = \overline{1}$. In $\mathbf{Z}_6$, each of the elements $\overline{2}$ and $\overline{3}$ do not have an inverse relative to •.

Theorem 6.4.7

Consider the set $\mathbf{Z}_p$, p prime, together with the operation •.

a. If $\overline{a}, \overline{b} \in \mathbf{Z}_p$ and $\overline{a} \bullet \overline{b} = \overline{0}$, then $\overline{a} = \overline{0}$ or $\overline{b} = \overline{0}$.

b. If $\overline{a} \in \mathbf{Z}_p$ such that $\overline{a} \neq \overline{0}$, then $\overline{a}$ has an inverse relative to •.

Proof

a. Let $\overline{a}, \overline{b} \in \mathbf{Z}_p$ such that $\overline{a} \bullet \overline{b} = \overline{0}$. Then $\overline{a \bullet b} = \overline{0}$ and $ab \equiv 0 \pmod{p}$. Therefore $p \mid (ab - 0)$ or $p \mid ab$. Since p is prime, $p \mid a$ or $p \mid b$. Hence $a \equiv 0 \pmod{p}$ or $b \equiv 0 \pmod{p}$. Thus $\overline{a} = \overline{0}$ or $\overline{b} = \overline{0}$.

■

b. Let $\overline{a} \in \mathbf{Z}_p$ where $\overline{a} \neq \overline{0}$. Now $\overline{a} \neq \overline{0}$ implies that $a \not\equiv 0 \pmod{p}$. Hence, $p \nmid (a - 0)$ or $p \nmid a$ and thus by Theorem 6.3.3, part (a), $1 = (a, p)$. By Theorem 6.4.2, $\exists x \in \mathbf{Z}$ such that $ax \equiv 1 \pmod{p}$. Thus $\overline{ax} = \overline{1}$ or $\overline{a} \bullet \overline{x} = \overline{1}$. But $\overline{x} = \overline{r}$ where $0 < r < p$. Hence, $\overline{a} \bullet \overline{r} = \overline{1}$. Since • is commutative, $\overline{r} \bullet \overline{a} = \overline{1}$. Therefore, $\overline{a}$ has an inverse relative to •.

■

$\mathbf{Z}_p$, where p is a positive prime integer, together with the operations + and • is an example of a mathematical system called a *field.*

Example 6.4.6

Consider the set $\mathbf{Z}_5 = \{\overline{0},\overline{1},\overline{2},\overline{3},\overline{4}\}$ together with the operations + and • .

+	$\overline{0}$	$\overline{1}$	$\overline{2}$	$\overline{3}$	$\overline{4}$
$\overline{0}$	$\overline{0}$	$\overline{1}$	$\overline{2}$	$\overline{3}$	$\overline{4}$
$\overline{1}$	$\overline{1}$	$\overline{2}$	$\overline{3}$	$\overline{4}$	$\overline{0}$
$\overline{2}$	$\overline{2}$	$\overline{3}$	$\overline{4}$	$\overline{0}$	$\overline{1}$
$\overline{3}$	$\overline{3}$	$\overline{4}$	$\overline{0}$	$\overline{1}$	$\overline{2}$
$\overline{4}$	$\overline{4}$	$\overline{0}$	$\overline{1}$	$\overline{2}$	$\overline{3}$

•	$\overline{0}$	$\overline{1}$	$\overline{2}$	$\overline{3}$	$\overline{4}$
$\overline{0}$	$\overline{0}$	$\overline{0}$	$\overline{0}$	$\overline{0}$	$\overline{0}$
$\overline{1}$	$\overline{0}$	$\overline{1}$	$\overline{2}$	$\overline{3}$	$\overline{4}$
$\overline{2}$	$\overline{0}$	$\overline{2}$	$\overline{4}$	$\overline{1}$	$\overline{3}$
$\overline{3}$	$\overline{0}$	$\overline{3}$	$\overline{1}$	$\overline{4}$	$\overline{2}$
$\overline{4}$	$\overline{0}$	$\overline{4}$	$\overline{3}$	$\overline{2}$	$\overline{1}$

Not only is $\mathbf{Z}_5$ together with the operations + and • a commutative ring with identity, but since 5 is a prime integer, $\mathbf{Z}_5$ possesses additional properties (see Theorem 6.4.7) and thus is also a field.

Exercises 6.4

1. Let $m = 2$ and $a, b \in \mathbf{Z}$.
   a. If a and b are both even or both odd, show $a \equiv b \pmod 2$.
   b. If a is even and b is odd, show $a \not\equiv b \pmod 2$.

2. Let $a, b, m \in \mathbf{Z}$, $m > 1$. If a and b have the same remainder when divided by m, show $a \equiv b \pmod m$.

3. Prove part (a) of Theorem 6.4.1.

4. Let $a, b, m \in \mathbf{Z}$ where $m > 1$ and $1 = (a, m)$. By Theorem 6.4.2, $\exists\, x \in \mathbf{Z}$ such that $ax \equiv b \pmod m$. Show $\forall t \in \overline{x}$, $at \equiv b \pmod m$. See Example 6.4.2, part (b).

5. Let $a, b, c, m \in \mathbf{Z}$ where $m > 1$. If $1 = (a, m)$ and $ab \equiv ac \pmod m$, show $b \equiv c \pmod m$.

6. Construct examples to show that the conclusions of Theorem 6.4.2 and problem (5) are not necessarily true if $1 \neq (a, m)$.

7. Prove parts (b), (c), (e) and (h) of Theorem 6.4.6.

8. Construct an + and • table for $\mathbf{Z}_6$.

Chapter 7

FINITE AND INFINITE SETS

In this chapter we shall discuss the idea of the "size" of a set. Our everyday experiences give us an understanding of finite sets and thus we will use this understanding to help define infinite sets.

Section 7.1
EQUIVALENT SETS

Two sets should have the same "size" if they have the same "number of elements." Using the concept of a function we can formalize this idea.

Definition 7.1.1

Two sets, A and B, are said to be ***equivalent*** (equipotent) and we write $\mathbf{A \sim B}$ if there exists a function $f: A \to B$ which is a 1–1 correspondence.

Example 7.1.1

a. Let $A = \{a,b,c,d\}$ and $\mathbf{N}_4 = \{1,2,3,4\}$. The function $f: A \to \mathbf{N}_4$ where $f(a) = 1$, $f(b) = 2$, $f(c) = 3$ and $f(d) = 4$ is a 1–1 correspondence. Hence $A \sim \mathbf{N}_4$.

b. Let $\mathbf{E} = \{x \in \mathbf{Z} \mid x = 2k, k \in \mathbf{Z}\}$, the set of even integers. The function $f: \mathbf{Z} \to \mathbf{E}$ where $f(k) = 2k$ is a 1–1 correspondence. Hence $\mathbf{Z} \sim \mathbf{E}$.

c. Consider the open interval $(0,1)$ and any other open interval, say (a,b). The function $f: (0,1) \to (a,b)$ where $f(x) = (b-a)x + a$ is a 1–1 correspondence. Hence, $(0,1) \sim (a,b)$. The function f is constructed as follows:

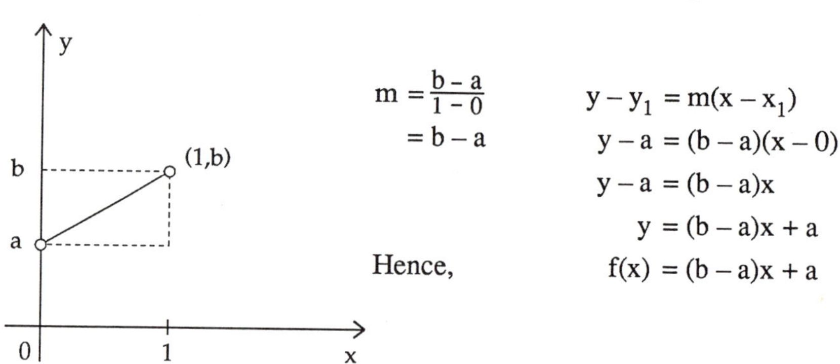

$$m = \frac{b-a}{1-0} = b - a$$

$$\begin{aligned} y - y_1 &= m(x - x_1) \\ y - a &= (b-a)(x-0) \\ y - a &= (b-a)x \\ y &= (b-a)x + a \end{aligned}$$

Hence, $f(x) = (b-a)x + a$

d. Let $\mathbf{R}$ be the set of real numbers. The function $f: (-\frac{\pi}{2}, \frac{\pi}{2}) \to \mathbf{R}$ where $f(x) = \tan x$ is a 1–1 correspondence. Hence $(-\frac{\pi}{2}, \frac{\pi}{2}) \sim \mathbf{R}$.

Theorem 7.1.1

Let A, B and C be sets.

a. If $A \sim B$, then $B \sim A$.

b. If $A \sim B$ and $B \sim C$, then $A \sim C$.

Proof

a. Let $A \sim B$. Then there exists a function $f: A \to B$ which is a 1–1 correspondence. Thus the function $f^{-1}: B \to A$ exists and is a 1–1 correspondence. Therefore, $B \sim A$.

■

b. Let $A \sim B$ and $B \sim C$. Then there exist functions $f: A \to B$ and $g: B \to C$ which are 1–1 correspondences. Thus the function $g \circ f: A \to C$ is a 1–1 correspondence. Therefore, $A \sim C$.

■

Theorem 7.1.2

If $A \sim B$ and $C \sim D$, then $A \times C \sim B \times D$.

Proof

Let $A \sim B$ and $C \sim D$. Then there exists functions $f: A \to B$ and $g: C \to D$ which are 1–1 correspondences. Let $h: A \times C \to B \times D$ where $h((a,c)) = (f(a), g(c))$. We shall show h is a 1–1 correspondence and hence, $A \times C \sim B \times D$.

Let $(a,c), (a',c') \in A \times C$. Suppose $h((a,c)) = h((a',c'))$. Then $(f(a),g(c)) = (f(a'),g(c'))$ which yields $f(a) = f(a')$ and $g(c) = g(c')$. Since f and g are both 1–1 functions, $a = a'$ and $c = c'$. Therefore, $(a,c) = (a',c')$ and thus h is 1–1.

Let $(b,d) \in B \times D$. We must show there exists $(x,y) \in A \times C$ such that $h((x,y)) = (b,d)$. Now $b \in B$ and f is onto implies $\exists a \in A$ such that $f(a) = b$. In like manner since $d \in D$ and g is onto, $\exists c \in C$ such that $g(c) = d$. Therefore, $(a,c) \in A \times C$, $h((a,c)) = (f(a),g(c)) = (b,d)$ and thus h is onto.

■

Exercises 7.1

1. Let $H = \{x \in \mathbf{Z} \mid x = 3k, k \in \mathbf{Z}\}$. Show $\mathbf{Z} \sim H$.

2. Let O be the set of odd natural numbers; i.e., $O = \{x \in \mathbf{N} \mid x = 2n - 1, n \in \mathbf{N}\}$. Show $\mathbf{N} \sim O$. Let E be the set of even natural numbers, i.e., $E = \{x \in \mathbf{N} \mid x = 2n, n \in \mathbf{N}\}$. Show $\mathbf{N} \sim E$.

3. Let $A \sim B$ and $C \sim D$. If $A \cap C = \emptyset$ and $B \cap D = \emptyset$, show $A \cup C \sim B \cup D$. (Hint: Construct a function $f \cup g: A \cup C \rightarrow B \cup D$.)

4. a. Let $X \neq \emptyset$ and consider the power set, P(X). For A, B $\in$ P(X), let $A \mathcal{R} B \Leftrightarrow A \sim B$. Show $\mathcal{R}$ is an equivalence relation on P(X).

   b. Let $X = \{a,b,c\}$. Find P(X). Since $\sim$ is an equivalence relation on P(X), $\sim$ induces a partition of P(X). Find the equivalence sets of P(X) determined by $\sim$. (Recall that the equivalence sets are the elements of the partition.)

Section 7.2
FINITE SETS

We intuitively think of a set as finite if we "can count its elements." Of course, this implies that we know when we have finished counting. The following definition formalizes this concept.

Definition 7.2.1

A set A is said to be ***finite*** if either $A = \emptyset$ or $A \sim \mathbf{N}_k = \{1,2,\ldots,k\}$ for some natural number k.

If $A \neq \emptyset$ is a finite set, then there exists a natural number k and a function $f: \mathbf{N}_k \rightarrow A$ which is a 1–1 correspondence. Thus, $A = f(\mathbf{N}_k) = \{f(n) \mid n \in \mathbf{N}_k\}$. Since $\mathbf{N}_k$ is an ordered set, we can use f to impose an order on A. By letting $f(n) = a_n$, the set A can be written in a more convenient form as $A = \{a_1,a_2,\ldots,a_k\}$.

Example 7.2.1

a. Refer to Example 7.1.1, part (a). Since $A = \{a,b,c,d\} \sim \mathbf{N}_4$, A is a finite set.

b. **N** is not a finite set. Suppose **N** is a finite set. Then there exists $k \in \mathbf{N}$ such that $\mathbf{N}_k \sim \mathbf{N}$ and we can then write **N** in the form $\mathbf{N} = \{x_1,x_2,\ldots,x_k\}$ where the x_i are distinct natural numbers. Now $x = x_1 + x_2 + \ldots + x_k + 1 \in \mathbf{N}$ and $x = x_1 + x_2 + \ldots + x_k + 1 \neq x_i$ for all i, $i = 1, 2, \ldots, k$, since $x > x_i$ for all i. Hence, $\mathbf{N} \neq \{x_1,x_2,\ldots,x_k\}$. Thus we have $\mathbf{N} = \{x_1,x_2,\ldots,x_k\}$ and $\mathbf{N} \neq \{x_1,x_2,\ldots,x_k\}$ which is a contradiction. This contradiction arose from assuming **N** is a finite set. Therefore, **N** is not a finite set.

Theorem 7.2.1

If A is a finite set, then $A \cup \{x\}$ is a finite set.

Proof

Let A be a finite set.

$A = \emptyset$: If $A = \emptyset$, then $A \cup \{x\} = \{x\}$.
Since $\{x\} \sim \mathbf{N}_1$, $A \cup \{x\} \sim \mathbf{N}_1$ and thus $A \cup \{x\}$ is a finite set.

$A \neq \emptyset$: If $A \neq \emptyset$ we have two cases to consider.

i. **$x \in A$:** If $x \in A$, then $A \cup \{x\} = A$ and thus $A \cup \{x\}$ is a finite set.

ii. **$x \notin A$:** Since A is finite, $\exists k \in \mathbf{N}$ such that $\mathbf{N}_k \sim A$. Hence, there exists a function $f: \mathbf{N}_k \rightarrow A$ which is a 1–1 correspondence such that $f(n) = a_n$, $n = 1, \ldots, k$. Since $A \subseteq A \cup \{x\}$, $f: \mathbf{N}_k \rightarrow A \cup \{x\}$. We wish to construct a function $g: \mathbf{N}_{k+1} \rightarrow A \cup \{x\}$ such that g is an extension of f and g is a 1–1 correspondence. Denote x by a_{k+1}. Then $A \cup \{x\} = \{a_1, \ldots, a_k, a_{k+1}\}$ and define $g: \mathbf{N}_{k+1} \rightarrow A \cup \{x\}$ by:

$$g(n) = \begin{cases} f(n) & \text{when } n \in \mathbf{N}_k \\ a_{k+1} & \text{when } n = k + 1. \end{cases}$$

Then g is a 1–1 correspondence, $\mathbf{N}_{k+1} \sim A \cup \{x\}$ and thus $A \cup \{x\}$ is a finite set. ■

Theorem 7.2.2

If A is a finite set, then $A \cup \{x_1,x_2, \ldots,x_n\}$ is a finite set.

Proof

Let A be a finite set; and, for each $n \in \mathbf{N}$, let p(n) be the statement:

$$A \cup \{x_1,x_2,\ldots,x_n\} \text{ is a finite set.}$$

p(1): Show p(1) is true; i.e., $A \cup \{x_1\}$ is a finite set.

Since A is finite, $A \cup \{x_1\}$ is a finite set by Theorem 7.2.1. Therefore, p(1) is true.

p(k): Assume p(k) is true for some k, $k \geq 1$. Then $A \cup \{x_1,x_2,\ldots,x_k\}$ is a finite set.

p(k + 1): Show p(k + 1) is true; i.e., $A \cup \{x_1,x_2,\ldots,x_k,x_{k+1}\}$ is a finite set.

Now $A \cup \{x_1,x_2,\ldots, x_k,x_{k+1}\} = A \cup (\{x_1,x_2,\ldots,x_k\} \cup \{x_{k+1}\})$
$= (A \cup \{x_1,x_2,\ldots,x_k\}) \cup \{x_{k+1}\}$.

Since p(k) is true, $A \cup \{x_1,x_2,\ldots,x_k\}$ is a finite set; and, by Theorem 7.2.1, $(A \cup \{x_1,x_2,\ldots,x_k\}) \cup \{x_{k+1}\}$ is also a finite set. Therefore, $A \cup \{x_1,x_2,\ldots,x_k,x_{k+1}\}$ is a finite set and p(k + 1) is true.

Therefore, $(\forall n)(p(n))$ is true.

■

Corollary

If A and B are finite sets, then $A \cup B$ is a finite set.

Note that this corollary states that the union of two finite sets is finite. The next theorem states that a finite union of finite sets is also finite.

Theorem 7.2.3

If $A_1, A_2, \ldots, A_n$ are finite sets, then $\bigcup_{i=1}^{n} A_i$ is a finite set.

Proof

For each $n \in \mathbf{N}$, $n \geq 2$, let p(n) be the statement:

If $A_1, A_2, \ldots, A_n$ are finite sets, then $\bigcup_{i=1}^{n} A_i$ is a finite set.

p(2): Show p(2) is true, ie., if A_1 and A_2 are finite sets, then $A_1 \cup A_2$ is a finite set.

Since A_1 and A_2 are finite sets, $A_1 \cup A_2$ is a finite set by the Corollary to Theorem 7.2.2. Therefore, p(2) is true.

p(k): Assume p(k) is true for some k, $k \geq 2$; i.e., if $A_1, A_2, \ldots, A_k$ are finite sets, then $\bigcup_{i=1}^{k} A_i$ is a finite set.

p(k + 1): Show p(k + 1) is true; i.e., if $A_1, A_2, \ldots, A_k, A_{k+1}$ are finite sets, then $\bigcup_{i=1}^{k+1} A_i$ is a finite set.

Now $\bigcup_{i=1}^{k+1} A_i = (\bigcup_{i=1}^{k} A_i) \cup A_{k+1}$. Since p(k) is true, $\bigcup_{i=1}^{k} A_i$ is a finite set; and, by the Corollary to Theorem 7.2.2, $(\bigcup_{i=1}^{k} A_i) \cup A_{k+1}$ is a finite set. Therefore, $\bigcup_{i=1}^{k+1} A_i$ is a finite set and p(k + 1) is true.

Therefore, $(\forall n, n \geq 2)\ (p(n))$ is true. ■

Theorem 7.2.4

If A is a finite set, then any subset of A is a finite set.

Proof

Let A be a finite set. We have two cases to consider.

$A = \emptyset$: If $B \subseteq A$, then $B = \emptyset$ and B is a finite set.

$A \neq \emptyset$: For each $n \in \mathbf{N}$, let p(n) be the statement:
If $A = \{a_1, a_2, \ldots, a_n\}$, then any subset of A is a finite set.

p(1): Show p(1) is true; i.e., if $A = \{a_1\}$ and $B \subseteq A$, then B is finite.

If $B \subseteq A$, then $B = \emptyset$ or $B = A$. Therefore, B is finite and p(1) is true.

p(k): Assume p(k) is true; i.e., if $A = \{a_1, a_2, \ldots, a_k\}$, then any subset of A is a finite set.

p(k + 1): Show p(k + 1) is true; i.e., if $A = \{a_1, a_2, \ldots, a_k, a_{k+1}\}$ and $B \subseteq A$, then B is finite.

Let $A = \{a_1, a_2, \ldots, a_k, a_{k+1}\}$ and $B \subseteq A$. We have two cases to consider.

i. **$a_{k+1} \notin B$:** If $a_{k+1} \notin B$, then $B \subseteq \{a_1, a_2, \ldots, a_k\}$; and, since p(k) is true, B is a finite set.

ii. **$a_{k+1} \in B$:** Let $C = B - \{a_{k+1}\}$. Then $C \subseteq \{a_1, a_2, \ldots, a_k\}$; and, p(k) is true, C is a finite set. By Theorem 7.2.1, $\{a_{k+1}\} = B$ is a finite set.

Therefore p(k + 1) is true.

Thus, $(\forall n)(p(n))$ is true.

■

Theorem 7.2.5

If A and B are finite sets, then A x B is a finite set.

Proof

Let A and B be finite sets.

If $A = \emptyset$ or $B = \emptyset$, then $A \times B = \emptyset$. Hence, A x B is finite.

If $A \neq \emptyset$ and $B \neq \emptyset$, then $A \times B \neq \emptyset$. Since A and B are finite, A and B can be represented as $A = \{a_1, \ldots, a_m\}$ and $B = \{b_1, \ldots, b_n\}$. Now A has m elements and B has n elements. Hence, by Theorem 3.2.1, A x B has mn elements. Therefore, $A \times B \sim N_{mn}$ and thus A x B is finite.

■

Exercises 7.2

1. If $A \sim B$ and A is a finite set, show that B is a finite set.

2. If A is a finite set and $f: A \rightarrow B$ such that f is onto, show that B is a finite set.

3. If $A_1, \ldots, A_n$ are finite sets, $n \geq 2$, use mathematical induction to prove the product set $A_1 \times \ldots \times A_n$ is a finite set. (Hint: Make use of the fact that $(A_1 \times \ldots \times A_n) \times A_{n+1} \sim A_1 \times \ldots \times A_1 \times A_{n+1}$. See Exercises 3.2, problem (6).)

Section 7.3
INFINITE SETS

In the previous section we defined a set A as being finite if either $A = \emptyset$; or, if $A \neq \emptyset$, then A is equivalent to the set $N_k = \{1,2, \ldots ,k\}$. Since we now have a clear idea of a finite set, we can use this concept to identify sets that are not finite.

Definition 7.3.1

A set A is said to be an ***infinite*** set if A is not a finite set.

Theorem 7.3.1

Let A be an infinite set.

a. If $A \sim B$, then B is an infinite.

b. If $A \subseteq B$, then B is an infinite set.

Proof
See Exercises 7.3, problem (1).

Example 7.3.1

a. The set of natural numbers, $\mathbf{N}$, is an infinite set (see Example 7.2.1, part (b)).

b. By Theorem 7.3.1, $\mathbf{Z}$, $\mathbf{Q}$, $\mathbf{R}$ are infinite sets since $\mathbf{N} \subseteq \mathbf{Z}$, $\mathbf{N} \subseteq \mathbf{Q}$ and $\mathbf{N} \subseteq \mathbf{R}$.

c. The open interval $(0,1) = \{x \in \mathbf{R} \mid 0 < x < 1\}$ is an infinite set. From Example 7.1.1., parts (c) and (d), we know $(0,1) \sim (-\frac{\pi}{2}, \frac{\pi}{2})$ and $(-\frac{\pi}{2}, \frac{\pi}{2}) \sim \mathbf{R}$. Hence by Theorem 7.1.1, $(0, 1) \sim \mathbf{R}$. Since $\mathbf{R}$ is infinite, (0,1) is infinite by Theorem 7.3.1, part (a).

Definition 7.3.1 is not the only way of defining an infinite set. In the late nineteenth century there was a great deal of controversy concerning how to define an infinite set. Another possibility was proposed by Richard Dedekind who defined a set A as being an infinite set if it was equivalent to a proper subset of itself. For example, this definition can be used to show that **N** is an infinite set since E, the set of even natural numbers, is a proper subset of **N** and **N** ~ E.
(See Exercises 7.1, problem (2).)

The set of natural numbers holds an important place in the theory of infinite sets. This is exhibited by the following definition.

Definition 7.3.2

A set A is said to be ***denumerable*** if A ~ **N**.

If A is a denumerable set, then there exists a function f: **N** → A which is a 1–1 correspondence. Thus A = f(**N**) = {f(n) | n ∈ **N**}. Since **N** is an ordered set, we can use f to impose an ordering on the set A. By letting f(n) = a_n, the set A can be written in a more convenient form as an infinite sequence of distinct elements:

$$A = \{a_1, a_2, \ldots, a_n, \ldots\}.$$

Example 7.3.2

a. Since **N** ~ **N**, **N** is a denumerable set.

b. The set $A = \{1, \frac{1}{2}, \frac{1}{3}, \frac{1}{4}, \ldots, \frac{1}{n}, \ldots\}$ is a denumerable set since the function f: **N** → A where $f(n) = \frac{1}{n}$ is a 1–1 correspondence.

c. The set of negative integers $\mathbf{Z}^- = \{-1, -2, -3, \ldots\}$ is a denumerable set since the function f: **N** → $\mathbf{Z}^-$ where f(n) = -n is a 1–1 correspondence.

Theorem 7.3.2

If A is a denumerable set, then $A \cup \{x\}$ is a denumerable set.

Proof

Let A be a denumerable set. We have two cases to consider.

$x \in A$: If $x \in A$, then $A \cup \{x\} = A$ and hence $A \cup \{x\}$ is a denumerable set.

$x \notin A$: Since $\mathbf{N} \sim A$, there exists $f: \mathbf{N} \to A$ which is a 1–1 correspondence such that $f(n) = a_n$. Let $x = a_0$. Then $A \cup \{x\} = \{a_0, a_1, \ldots, a_n, \ldots\}$ and the function $g: \mathbf{N} \to A \cup \{x\}$ defined by

$$g(n) = a_{n-1} = \begin{cases} a_o & \text{when } n = 1 \\ f(n-1) & \text{when } n > 1 \end{cases}$$

is a 1–1 correspondence. Therefore, $N \sim A \cup \{x\}$ and $A \cup \{x\}$ is a denumerable set.

■

Theorem 7.3.3

If A is a denumerable set, then $A \cup \{x_1, x_2, \ldots, x_n\}$ is a denumerable set.

Proof

See Exercises 7.3, problem (2).

Corollary

If A is a denumerable set, and B is a finite set, then $A \cup B$ is a denumerable set.

Theorem 7.3.4

If A is a denumerable set and if $B \subseteq A$ such that B is infinite, then B is a denumerable set.

Proof

Let A be a denumerable set and B ⊆ A such that B is infinite. Now A may be represented as $A = \{a_1, a_2, \ldots, a_n, \ldots\}$. Let

a_{n_1} be the 1st element in A for which $a_{n_1} \in B$
a_{n_2} be the 2nd element in A for which $a_{n_2} \in B$
•   •
•   •
•   •
a_{n_k} be the kth element in A for which $a_{n_k} \in B$

Since A, B and $A - \{a_{n_1}, \ldots, a_{n_k}\}$ are infinite sets (see Exercises 7.3, (5)), the process can be continued. Hence, $B = \{a_{n_1}, a_{n_2}, \ldots, a_{n_k}, \ldots\}$ and the function $f: \mathbf{N} \to B$ where $f(k) = a_{n_k}$ is 1–1 and onto and thus a 1–1 correspondence. Therefore, $\mathbf{N} \sim B$ and B is denumerable.

■

Example 7.3.3

In Example 3.2.1, part (d), we showed the function $h: \mathbf{N} \times \mathbf{N} \to \mathbf{N}$ where $h((m,n)) = 2^m(2n-1)$ is 1–1. We wish to show **N** x **N** is denumerable. Since $h: \mathbf{N} \times \mathbf{N} \to \mathbf{N}$ is 1–1, $h: \mathbf{N} \times \mathbf{N} \to R_h = h(\mathbf{N} \times \mathbf{N})$ is a 1–1 correspondence. Hence $\mathbf{N} \times \mathbf{N} \sim h(\mathbf{N} \times \mathbf{N})$.

The function $f: \mathbf{N} \to \mathbf{N} \times \{1\}$ where $f(n) = (n,1)$ is a 1–1 correspondence. Thus, $\mathbf{N} \sim \mathbf{N} \times \{1\}$ and $\mathbf{N} \times \{1\}$ is infinite. Since $\mathbf{N} \times \{1\} \subseteq \mathbf{N} \times \mathbf{N}$, **N** x **N** is infinite and hence so is h(**N** x **N**).

Now $h(\mathbf{N} \times \mathbf{N}) \subseteq E \subseteq \mathbf{N}$ and thus h(**N** x **N**) is denumerable by Theorem 7.3.4. Consequently, we have $\mathbf{N} \times \mathbf{N} \sim h(\mathbf{N} \times \mathbf{N})$ and $h(\mathbf{N} \times \mathbf{N}) \sim \mathbf{N}$ which, by Theorem 7.1.1, part (b), yields $\mathbf{N} \times \mathbf{N} \sim \mathbf{N}$. Therefore, **N** x **N** is denumerable.

We already know that the union of two finite sets is finite. The next result shows that the union of two denumerable sets is also denumerable.

Theorem 7.3.5

If A and B are denumerable sets, then $A \cup B$ is a denumerable set.

Proof

Let A and B be denumerable sets. We have two cases to consider.

$\mathbf{A \cap B = \emptyset}$: A and B denumerable implies $A \sim \mathbf{N}$ and $B \sim \mathbf{N}$. If $E = \{2n \mid n \in \mathbf{N}\}$ is the set of even natural numbers and $0 = \{2n - 1 \mid n \in \mathbf{N}\}$ the set of odd natural numbers, then $\mathbf{N} \sim E$ and $\mathbf{N} \sim O$. (See, Exercises 7.1, problem (2).) Since $A \sim \mathbf{N}$ and $\mathbf{N} \sim E$, $A \sim E$; and, since $B \sim \mathbf{N}$ and $\mathbf{N} \sim O$, $B \sim O$. Hence, there exists functions $f: A \rightarrow E$ and $g: B \rightarrow O$ which are 1–1 correspondences. Since $A \cap B = \emptyset$, define the function $F: A \cup B \rightarrow E \cup O = \mathbf{N}$ as follows:

$$F(x) = (f \cup g)(x) = \begin{cases} f(x) & \text{when } x \in A \\ g(x) & \text{when } x \in B \end{cases}$$

Then F is 1–1 and onto and consequently $A \cup B$ is denumerable.

$\mathbf{A \cap B \neq \emptyset}$: Now $A \cup B = A \cup (B - A)$ (see Exercises 2.3., problem (8), part (b)).

If $B - A$ is finite, then by Corollary to Theorem 7.3.3, $A \cup (B - A) = A \cup B$ is a denumerable set.

If $B - A$ is infinite, then by Theorem 7.3.4, $B - A$ is denumerable since B is denumerable and $B - A \subseteq B$. Now $A \cap (B - A) = \emptyset$ (see Exercise 2.3, problem (8), part (a)). By the first case, $A \cup (B - A) = A \cup B$ is denumerable.

■

We can also generalize this result to state that the finite union of denumerable sets is also denumerable.

Theorem 7.3.6

If $A_1, A_2, \ldots, A_n$ are denumerable sets, then $\bigcup_{i=1}^{n} A_i$ is a denumerable set.

Proof

See Exercises 7.3, problem (6).

Example 7.3.4

a. $\mathbf{Z} = \mathbf{Z}^- \cup \{0\} \cup \mathbf{Z}^+$ is a denumerable set. f: $\mathbf{N} \to \mathbf{Z}^-$ where f(n) = -n and g: $\mathbf{N} \to \mathbf{Z}^+$ where g(n) = n are 1–1 correspondences. Hence, $\mathbf{N} \sim \mathbf{Z}^-$ and $\mathbf{N} \sim \mathbf{Z}^+$. Thus $\mathbf{Z}^-$ and $\mathbf{Z}^+$ are denumerable sets. Therefore, by Theorem 7.3.2 and Theorem 7.3.5, **Z** is a denumerable set.

b. $\mathbf{Q} = \mathbf{Q}^- \cup \{0\} \cup \mathbf{Q}^+$ is a denumerable set. f: $\mathbf{N} \times \mathbf{N} \to \mathbf{Q}^-$ where $f((m,n)) = -\frac{m}{n}$ and g: $\mathbf{N} \times \mathbf{N} \to \mathbf{Q}^+$ where $g((m,n)) = \frac{m}{n}$ are 1–1 correspondences. Since $\mathbf{N} \sim \mathbf{N} \times \mathbf{N}$, $\mathbf{N} \sim \mathbf{Q}^-$ and $\mathbf{N} \sim \mathbf{Q}^+$. Thus $\mathbf{Q}^-$ and $\mathbf{Q}^+$ are denumerable sets. Therefore, by Theorem 7.3.2 and Theorem 7.3.5, **Q** is a denumerable set.

Theorem 7.3.7

If A and B are denumerable sets, then A x B is a denumerable set.

Proof

Let A and B be denumerable sets. Then A ~ N and B ~ N. By Theorem 7.1.2, A x B ~ N x N. Since N x N ~ N, A x B ~ N. Therefore, A x B is a denumerable set. ∎

Exercises 7.3

1. Prove Theorem 7.3.1. (Hint: Use the method of contradiction and theorems from Section 7.2.)

2. Prove Theorem 7.3.3. (Hint: See proof of Theorem 7.2.2.)

3. Let A ~ B. Show that if A is denumerable, then B is denumerable.

4. If A is an infinite set, show that $A - \{a_1, a_2, \ldots, a_n\}$ where the $a_i \in A$ is an infinite set.

5. Show that every infinite set contains a denumerable subset. (Hint: Use result of problem (4) to form a denumerable set.)

6. Prove Theorem 7.3.6. (Hint: See proof of Theorem 7.2.3.)

7. If A is denumerable and B is finite, show A x B is denumerable. (Hint: For each $b_i \in B$, consider the set $A \times \{b_i\}$.)

Section 7.4
COUNTABLE SETS

There are infinite sets which are not denumerable; i.e., infinite sets for which there does not exist a 1–1 correspondence with the set of natural numbers **N**. Hence, there are different types of infinite sets. In this section we shall look at sets that are *countably* infinite and *uncountably* infinite.

Definition 7.4.1

A set A is said to be ***countable*** if A is either a finite set or a denumerable set.

Example 7.4.1

a. The set A = {1,2,3,4} is countable since A is a finite set.

b. The sets **N**, **Z** and **Q** are countable since they are denumerable sets.

Theorem 7.4.1

If A is a countable set and if B ⊆ A, then B is a countable set.

Proof

Let A be a countable set and B ⊆ A. We have two cases to consider.

A finite: By Theorem 7.2.4, B is finite and hence countable.

A denumerable: If B is finite, then B is countable. If B is infinite, then by Theorem 7.3.4, B is denumerable and hence countable.

■

Thus we see subsets of countable sets are also countable. Using the results of the previous sections, we can also establish that union of two countable sets is also countable. Finally, we shall state analogous properties for the finite union of countable sets and the product set of two countable sets.

Theorem 7.4.2

If A and B are countable sets, than $A \cup B$ is a countable set.

> Proof
>
> Let A and B be two countable sets. We have three cases to consider.
>
> **A and B finite**: By Corollary to Theorem 7.2.2, $A \cup B$ is a finite set and hence countable.
>
> **A denumerable and B finite**: By Corollary to Theorem 7.3.3, $A \cup B$ is a denumerable set and hence countable.
>
> **A and B are denumerable**: By Theorem 7.3.5, $A \cup B$ is a denumerable set and hence countable.
>
> ■

Theorem 7.4.3

If $A_1, A_2, \ldots, A_n$ are countable sets, then $\bigcup_{i=1}^{n} A_i$ is a countable set.

> Proof
>
> See Exercises 7.4, problem (1).

Theorem 7.4.4

If A and B are countable sets, then A x B is a countable set.

> Proof
>
> See Exercises 7.4, problem (2).

Essentially, a non-empty set is countable if it is equivalent to **N** or a subset of **N**. We have shown that the integers, **Z**, and the rational numbers, **Q**, are countably infinite sets. Is the set of real numbers, **R**, a countable set? To answer this question, we shall show that the open interval $(0,1) = \{x \in \mathbf{R} \mid 0 < x < 1\}$ is not a countable set and thus, by Theorem 7.4.1, **R** is not a countable set. Therefore, **R** and any open interval $(a,b) = \{x \in \mathbf{R} \mid a < x < b\}$ are examples of uncountably infinite sets.

Theorem 7.4.5

The set $A = (0,1)$ is not a countable set.

Proof

> **Discussion**
> Any real number $x \in A = (0,1)$ can be represented in decimal form by an infinite string of digits; i.e., $x = 0.x_1x_2x_3 \ldots$ where $x_i \in \{0,1,2, \ldots, 9\}$. In order to guarantee the uniqueness of representation of each number $x \in (0,1)$, we shall agree that no decimal representation shall end in an infinite string of nines. Thus, if $x = \frac{27}{40}$, the decimal representation of x will be written as $0.675000\ldots$ and not as $0.674999\ldots$

Suppose $A = (0,1)$ is a countable set. Since $B = \{\frac{1}{2}, \frac{1}{3}, \frac{1}{4}, \ldots\} \subseteq A$ and since $B \sim \mathbf{N}$, by Theorem 7.3.1, part (b), A is an infinite set. Thus, if A is countable, then $\mathbf{N} \sim A$ and there exists $f: \mathbf{N} \to A$ such that f is a 1–1 correspondence. Hence, A can be represented by $A = \{a_1, a_2, \ldots, a_n, \ldots\}$. Now $a_i \in A = (0,1)$ means each a_i can be written in decimal form as $a_i = 0.x_{i1}\, x_{i2} \ldots x_{ik} \ldots$ where $x_{ij} \in \{0,1,2, \ldots, 9\}$. Let $b = 0.b_1\, b_2 \ldots b_k \ldots$ where $b_i \in \{1,2, \ldots, 8\}$ and $b_1 \neq x_{11}, b_2 \neq x_{22}, \ldots, b_k \neq x_{kk}, \ldots$ Then $b \in A$ and $b \neq a_i\ \forall i \in \mathbf{N}$. Thus, there does not exist $n \in \mathbf{N}$ such that $f(n) = b$ which means that f is not onto and hence $\mathbf{N} \nsim A$. Consequently we have $\mathbf{N} \sim A$ and $\mathbf{N} \nsim A$ which is a contradiction. This contradiction arose from assuming $A = (0,1)$ is a countable set. Therefore, A is not countable.

■

The *cardinal number* of a set A, denoted by **#(A)**, may be thought of intuitively as a measure of the "size" of the set A or as the "number of elements" in the set A. Two sets, A and B, are said to have the same cardinality if and only if $A \sim B$. Thus if

$A = \emptyset$,	then $\#(A) = 0$	
$A \sim \mathbf{N}_k$,	then $\#(A) = k$	
$A \sim \mathbf{N}$,	then $\#(A) = \aleph_o$	(aleph-null)
$A \sim R$,	then $\#(A) = c$	(continuum)

Example 7.4.2

a. If A = {a,b,c,d}, then A ~ $\mathbf{N}_4$ and #(A) = 4.

b. Since **Z** ~ **N** and **Q** ~ **N**, #(**Z**) = #(**Q**) = $\aleph_o$.

c. Since (0,1) ~ $(-\frac{\pi}{2}, \frac{\pi}{2})$ and $(-\frac{\pi}{2}, \frac{\pi}{2})$ ~ **R**, (0,1) ~ **R**.
Hence, #(0,1) = c.

Exercises 7.4

1. Prove Theorem 7.4.3.
(See proof of Theorem 7.2.3.)

2. Prove Theorem 7.4.4. (You must consider three cases: A and B finite; A denumerable and B finite; and A and B denumerable.)

Chapter 8

THE REAL NUMBER SYSTEM

Throughout the previous chapters we have assumed a basic knowledge with respect to the operations of *addition*, +, and *multiplication*, • , as well as the order relation *less than or equal to*, $\leq$, as defined on the set of real numbers. In this chapter we shall review a few of the basic algebraic properties of **R** and discuss other facts which will be useful in subsequent courses in mathematics.

Section 8.1
THE ALGEBRAIC STRUCTURE OF R

The set of real numbers **R** together with the operations + and • forms a mathematical system called a *field* and thus satisfies the following familiar properties:

Addition

$a + b = b + a$	$\forall a, b \in \mathbf{R}$	(commutative law)
$a + (b + c) = (a + b) + c$	$\forall a, b, c \in \mathbf{R}$	(associative law)
$a + 0 = a$	$\forall a \in \mathbf{R}$	(additive identity)
$\forall a \in \mathbf{R}\ \exists\ -a \in \mathbf{R}$ such that	$a + (-a) = 0$	(additive inverse)

Multiplication

$ab = ba$	$\forall a, b \in \mathbf{R}$	(commutative law)
$a(bc) = (ab)c$	$\forall a, b, c \in \mathbf{R}$	(associative law)
$a1 = a$	$\forall a \in \mathbf{R}$	(multiplicative identity)
$\forall a \in \mathbf{R}, a \neq 0, \exists a^{-1} = \frac{1}{a} \in \mathbf{R}$ such that $a\,a^{-1} = 1$		(multiplicative inverse)

Addition and Multiplication

$a(b + c) = ab + ac$	$\forall a, b, c \in \mathbf{R}$	(distributive law)

In Chapter 4 we introduced the idea of a mathematical relation, $\mathcal{R}$, defined on a fixed set A together with the various properties $\mathcal{R}$ might possess — reflexive, symmetric, transitive. Of particular importance was a relation $\mathcal{R}$ that possessed all three properties; i.e., an equivalence relation, which enabled us to identify certain elements of A that were related under $\mathcal{R}$ as being equivalent. We now wish to discuss another important type of mathematical relation called a *partial order*. As the name suggests, a partial order $\mathcal{R}$ defined on a set A imposes a certain order or rank between two elements of A that are related by designating which element is "larger."

Definition 8.1.1

Let $\mathcal{R}$ be a relation defined on A.

a. $\mathcal{R}$ is ***anti-symmetric*** if, whenever a $\mathcal{R}$ b and b $\mathcal{R}$ a, then a = b.

$$(a,b) \in \mathcal{R} \wedge (b,a) \in \mathcal{R} \Rightarrow a = b$$

b. $\mathcal{R}$ is called a ***partial order*** on A if $\mathcal{R}$ is reflexive, anti-symmetric and transitive.

c. Set A together with a partial order $\mathcal{R}$ is called a ***partially ordered set***.

Example 8.1.1

Let **R** be the set of real numbers; and, for a, b ∈ **R**, define **a ≤ b** (a less than or equal to b) if b – a is a non-negative real number. Then ≤ is a partial order on **R**. We must show that ≤ is reflective, anti-symmetric and transitive.

i. Let a ∈ **R**. Then a – a = 0 and hence a ≤ a. Therefore ≤ is reflexive.

ii. Let a, b ∈ R such that a ≤ b and b ≤ a. Then b – a and a – b are non-negative real numbers. Suppose a ≠ b. Then b – a and a – b are positive real numbers; and, since the positive reals are closed under addition, (b – a) + (a – b) is also a positive real number. But (b – a) + (a – b) = 0. This is a contradiction arrived at by assuming a ≠ b. Therefore, a = b and ≤ is anti-symmetric.

iii. Let a, b, c ∈ **R** such that a ≤ b and b ≤ c. Then b – a and c – b are non-negative real numbers and hence so is (b – a) + (c – b). But (b – a) + (c – b) = c – a. Therefore, a ≤ c and ≤ is transitive.

Since ≤ is a partial order defined on **R**, **R** is a partially ordered set relative to ≤. When dealing with **R** together with the partial order relation ≤, we shall use the following additional notation: For a, b ∈ **R**,

a < b (a less than b) ⇔ a ≤ b and a ≠ b.
b ≥ a (b greater than or equal to a) ⇔ a ≤ b.

The following properties hold in **R** relative to ≤: For a, b, c ∈ **R**,

if a ≤ b, then a + c ≤ b + c;
if a ≤ b and c is positive, then ac ≤ bc;
if a ≤ b and c is negative, then ac ≥ bc.

Example 8.1.1 continued next page.

Example 8.1.1 Continued

Note that for any a, b $\in$ **R**, either b – a is non-negative or a – b is non-negative. Hence, either $a \le b$ or $b \le a$. That is, any two real numbers are comparable with respect to the partial order $\le$. Furthermore, if $a \ne b$, then $a < b$ or $b < a$. As a consequence, we have what is called the *Trichotomy Law*: For a, b $\in$ **R**, either $a < b$, $a = b$ or $b < a$.

Definition 8.1.2

Let $\mathcal{R}$ be a partial order defined on A.

a. If for all a, b $\in$ A, either a $\mathcal{R}$ b or b $\mathcal{R}$ a, then $\mathcal{R}$ is called a ***total order*** on A.

b. Set A together with a total order $\mathcal{R}$ is called a ***totally ordered set***.

From what was said above, we see that **R** is a totally ordered set relative to $\le$; and, since **R** is also a field relative to + and •, **R** is referred to as a *totally ordered field.*

Exercises 8.1

1. For a, b $\in$ **N**, let a $\mathcal{R}$ b $\Leftrightarrow$ a | b (a | b, a divides b, if there exists c $\in$ **N** such that b = ac). Show $\mathcal{R}$ is a partial order on **N**. Is $\mathcal{R}$ a total order on **N**?

2. Let $\mathbf{Z}^* = \mathbf{Z} - \{0\}$. For a, b $\in \mathbf{Z}^*$, let a $\mathcal{R}$ b $\Leftrightarrow$ a | b. Show $\mathcal{R}$ is not a partial order on $\mathbf{Z}^*$ (see Theorem 6.1.3).

3. Let $X \ne \emptyset$. For A, B $\in$ P(X), let A $\mathcal{R}$ B $\Leftrightarrow A \subseteq B$. Show $\mathcal{R}$ is a partial order on P(X). Is $\mathcal{R}$ a total order on P(X)?

4. Let $\mathcal{R}$ be a partial order defined on A. Show $\mathcal{R}^{-1}$ is also a partial order on A.

5. Show the following properties hold in **R** relative to $\le$: For a, b, c $\in$ **R**,
   a. if $a \le b$, then $a + c \le b + c$;
   b. if $a \le b$ and c is positive, then $ac \le bc$;
   c. if $a \le b$ and c is negative, then $ac \ge bc$.

6. If a, b $\in$ **R** and $a < b$, show the inequality $a < \frac{a+b}{2} < b$ holds. (Hint: Use results of problem (5).)

Section 8.2
THE COMPLETENESS OF R

The set of rational numbers, **Q**, like the real numbers, **R**, is also a totally ordered field. However, there are two important characteristics which differentiate **R** from **Q**. One such characteristic was given in Chapter 7 where it was shown that **Q** is a countable set while **R** is not countable. Hence, since $\mathbf{Q} \subseteq \mathbf{R}$ and while both sets are infinite, there are more real numbers than there are rational numbers (see Section 7.4: $\#(\mathbf{Q}) = \aleph_0$ and $\#(\mathbf{R}) = \mathbf{c}$). In this section we shall discuss yet another differentiable property between **Q** and **R**; i.e., the idea of completeness.

Definition 8.2.1

Let $A \subseteq \mathbf{R}$, $A \neq \emptyset$ and $x \in \mathbf{R}$.

a. The number x is called an ***upper bound*** of A if $\forall a \in A$, $a \leq x$.
b. The number x is called a ***lower bound*** of A if $\forall a \in A$, $x \leq a$.

Example 8.2.1

a. If $A = \{1, \frac{1}{2}, \frac{1}{3}, \frac{1}{4}, \ldots\}$, then 1 and any real number x for which $1 < x$ is an upper bound of A while 0 and any real number x for which $x < 0$ is a lower bound of A.

b. If $A = (-1,\infty) = \{x \in \mathbf{R} \mid -1 < x\}$, then -1 and any real number x for which $x < -1$ is a lower bound of A. A has no upper bound.

c. If $A = [0,2] = \{x \in \mathbf{R} \mid 0 \leq x \leq 2\}$, then 2 and any real number x for which $2 < x$ is an upper bound of A while 0 and any real number x for which $x < 0$ is a lower bound of A.

Definition 8.2.2

Let $A \subseteq \mathbf{R}$, $A \neq \emptyset$.

a. Then $u \in \mathbf{R}$ is called a ***least upper bound*** of A and we write $\boldsymbol{u} = \mathbf{lub\ A}$ if:
   i. u is an upper bound of A;
   ii. x is an upper bound of A, then $u \leq x$.

b. Then $l \in \mathbf{R}$ is called a ***greatest lower bound*** of A and we write $\boldsymbol{l} = \mathbf{glb\ A}$ if:
   i. l is a lower bound of A;
   ii. x is a lower bound of A, then $x \leq l$.

Another way of saying that $u = \text{lub A}$ is that u is the smallest of all the upper bounds of A. In like manner, if $l = \text{glb A}$, then l is the largest of all the lower bounds of A.

Theorem 8.2.1

Let $A \subseteq \mathbf{R}$, $A \neq \emptyset$.

a. If $u = \text{lub A}$, then u is unique.

b. If $l = \text{glb A}$, then l is unique.

Proof
We shall prove part (a) and leave the proof of part (b) as an exercise.

a. Let $u = \text{lub A}$ and $u' = \text{lub A}$. Now u' is an upper bound of A; and, since $u = \text{lub A}$, $u \leq u'$. Also u is an upper bound of A; and, since $u' = \text{lub A}$, $u' \leq u$. Therefore, since $\leq$ is anti-symmetric, $u = u'$. Thus, if $u = \text{lub A}$, then u is unique.

■

Example 8.2.2

Refer to Example 8.2.1.

a. In part (a) where $A = \{1, \frac{1}{2}, \frac{1}{3}, \frac{1}{4}, \ldots\}$ we have 1 = lub A and 0 = glb A.

1 = lub A: Now 1 is an upper bound of A. Let $x \in \mathbf{R}$ be an upper bound of A. Then $\forall a \in A$, $a \le x$. But $1 \in A$. Hence, $1 \le x$ and 1 = lub A.

0 = glb A: Now 0 is a lower bound of A. Let $x \in \mathbf{R}$ be a lower bound of A. Suppose $0 < x$. Then $\exists n \in \mathbf{N}$ such that $\frac{1}{n} < x$ (see Exercises 8.3, problem (1)). Hence, we have $\frac{1}{n} \in A$ and $0 < \frac{1}{n} < x$. Thus x is not a lower bound of A. This is a contradiction arrived at by assuming $0 < x$. Therefore, $x \le 0$ and 0 = glb A.

b. In part (b) where $A = (-1,\infty)$ we have -1 = glb A. Now -1 is a lower bound of A. Let x be a lower bound of A and suppose $-1 < x$. Then $\exists a \in \mathbf{R}$ such that $-1 < a < x$ (see Exercises 8.1, problem (6)). Hence, we have $a \in A$ and $a < x$. Thus x is not a lower bound of A. This is a contradiction arrived at by assuming $-1 < x$. Therefore, $x \le -1$ and -1 = glb A.

As mentioned in Chapter 1, certain mathematical statements are referred to as axioms; i.e., primary statements concerning one or more of the primitive terms. They specify properties and relations about the primitive terms and are accepted without question as being true. We shall accept the follow statement as an axiom in the ordered field of real numbers.

Axiom

Axiom of Completeness: Let $A \subseteq \mathbf{R}$, $A \neq \emptyset$. If A has an upper bound in **R**, then A has a least upper bound in **R**.

The above axiom gives another important characteristic of the real numbers. Recall there is a 1–1 correspondence between **R** and the set of points on a line. Hence, the real numbers can be represented geometrically as points on a line. From an intuitive point-of-view, the axiom tells us there are no "gaps" or "breaks" in the real line. The real line is "connected." For this reason, we say **R** is *complete*.

Let $a \in \mathbf{R}$ and consider the set $\mathbf{R} - \{a\}$. We shall show $\mathbf{R} - \{a\}$ is no longer complete since it has a "gap" and hence $\mathbf{R} - \{a\}$ is no longer "connected."

Intuitively, if we represent **R** – {a} geometrically as points on a line, the line has a "gap" or "break" at the point representing the number a.

Let $A = \{x \in \mathbf{R} - \{a\} \mid x < a\}$. Then $A \subseteq \mathbf{R} - \{a\}$ and $A \neq \emptyset$ since $(a - 1) \in A$. Now A has an upper bound in **R** – {a}; i.e., a + 1. However, A does not have a least upper bound in **R** – {a}; for suppose $u \in \mathbf{R} - \{a\}$ and $u = \text{lub } A$. There are two cases we must consider.

$u < a$: If $u < a$, then $u < \frac{u+a}{2} < a$. Since $\frac{u+a}{2} < a$, $\frac{u+a}{2} \in A$. Hence, u is not an upper bound of A and thus $u \neq \text{lub } A$. Contradiction.

$a < u$: If $a < u$, then $a < \frac{a+u}{2} < u$. Since $a < \frac{a+u}{2}$, $\frac{a+u}{2} \in \mathbf{R} - \{a\}$ and thus $\frac{a+u}{2}$ is an upper bound of A. Thus $u \neq \text{lub } A$. Contradiction.

In each case the contradiction arose by assuming $u = \text{lub } A$. Therefore, no such u exists.

Theorem 8.2.2

Let $A \subseteq \mathbf{R}$, $A \neq \emptyset$. If A has a lower bound in **R**, then A has a greatest lower bound in **R**.

Proof

Exercises 8.2, problem (5).

Example 8.2.3

Let $A = \{\frac{1-n}{2n+1} \in \mathbf{R} \mid n \in \mathbf{N}\}$. Determine if A has a lub and/or glb in **R**.

$A = \{0, -\frac{1}{5}, -\frac{2}{7}, -\frac{3}{9}, \ldots\}$ and the numbers in A are decreasing. Now $\forall a \in A$, $a \leq 0$.

Therefore, A has an upper bound in **R** and thus a lub in **R** ($0 = \text{lub } A$). Next, consider $\lim_{n \to \infty} [\frac{1-n}{2n+1}] = \lim_{n \to \infty} [\frac{\frac{1}{n} - 1}{2 + \frac{1}{n}}] = -\frac{1}{2}$. Now $\forall a \in A$, $-\frac{1}{2} \leq a$. Therefore, A has a lower bound in **R** and thus a glb in **R** ($-\frac{1}{2} = \text{glb } A$).

Exercises 8.2

1. Prove part (b) of Theorem 8.2.1.
2. Let $A = \{\frac{n}{n+1} \in \mathbf{R} \mid n \in \mathbf{N}\}$. Determine if A has a lub and/or glb in **R**.
3. Let $A \subseteq \mathbf{R}$, $A \neq \emptyset$.
   a. If $u = \text{lub } A$, show that $\forall\, \delta > 0\ \exists\, a \in A$ such that $u - \delta < a < u$.
   b. If $l = \text{glb } A$, show that $\forall\, \delta > 0\ \exists\, a \in A$ such that $l < a < l + \delta$.
4. Let $A \subseteq \mathbf{R}$, $A \neq \emptyset$. If $u = \text{lub } A$ and if $B = \{-x \in \mathbf{R} \mid x \in A\}$, show $-u = \text{glb } B$.
5. Prove Theorem 8.2.2. (Hint: Use problem (4).)
6. Let $A \subseteq B \subseteq \mathbf{R}$, $A \neq \emptyset$.
   a. If A and B have upper bounds in **R**, show lub A $\leq$ lub B.
   b. If A and B have lower bounds in **R**, show glb B $\leq$ glb A.

Section 8.3
FURTHER PROPERTIES OF R

In this section we shall introduce three important properties — the *Archimedean Property*, the *Dedekind Property* and the *Nested Interval Property* — each of which is a direct consequence of the Axiom of Completeness. The Archimedean Property will enable us to show that the rational numbers as well as the irrational numbers are *dense* in **R**. This latter result will in turn be used in Exercises 8.3, problem (3) to show that the rational numbers, **Q**, do not possess the completeness property.

Theorem 8.3.1 (Archimedean Property)

If a, b $\in \mathbf{R}^+$, then there exists n $\in \mathbf{N}$ such that b < na.

Proof

Let a, b $\in \mathbf{R}^+$ and suppose there does not exist n $\in \mathbf{N}$ such that b < na. Then $\forall n \in \mathbf{N}$, na $\le$ b. If A = {na | n $\in \mathbf{N}$}, then b is an upper bound of A. By the Axiom of Completeness, A has a lub, say u; and, $\forall n \in \mathbf{N}$, na $\le$ u. Let k $\in \mathbf{N}$. Then (k + 1) $\in \mathbf{N}$ and (k + 1)a $\in$ A. Now (k + 1)a $\le$ u and thus we have ka + a $\le$ u or ka $\le$ u – a. Since k is an arbitrary natural number, u – a is an upper bound of A. However, a $\in \mathbf{R}^+$ implies that u – a < u. Therefore, u $\ne$ lub A. This is a contradiction arrived at by assuming that there does not exist n $\in \mathbf{N}$ such that b < na. Hence, $\exists n \in \mathbf{N}$ such that b < na.

■

Theorem 8.3.2

If a, b $\in \mathbf{R}$ and a < b, then there exists a rational number r such that a < r < b.

Proof

Let a, b $\in \mathbf{R}$ such that a < b. Now 1, b – a $\in \mathbf{R}^+$ and hence by Theorem 8.3.1 $\exists n \in \mathbf{N}$ such that 1 < n(b – a). Thus 1 < nb – na or na + 1 < nb. Let A = {k $\in \mathbf{N}$ | na < k}. If na $\le$ 0, then 1 $\in$ A. If 0 < na, then na $\in \mathbf{R}^+$; and since 1 $\in \mathbf{R}^+$, $\exists k \in \mathbf{N}$ such that na < k • 1 = k and k $\in$ A. Hence A $\ne \emptyset$. Since $\mathbf{N} = \mathbf{Z}^+$ is a well-ordered set and A $\subseteq \mathbf{N}$, A has a smallest element m. Now m is the smallest positive integer such that na < m which implies (m – 1) $\le$ na or m $\le$ na + 1. We therefore have na < m $\le$ na + 1. From above, na + 1 < nb. Hence, na < m < nb which yields a < $\frac{m}{n}$ < b. Let r = $\frac{m}{n}$. Then r $\in \mathbf{Q}$ and a < r < b.

■

Theorem 8.3.3

If p is a positive prime integer, then $\sqrt{p}$ is an irrational number.

Proof

Let p be a positive prime integer and suppose $\sqrt{p}$ is a rational number. Then $\sqrt{p} = \frac{m}{n}$ where $m, n \in \mathbf{N}$ and $(m,n) = 1$.

$$\sqrt{p} = \frac{m}{n}$$
$$(\sqrt{p})n = m$$
$$pn^2 = m^2$$

Now $p \mid m^2$; and, since p is prime, $p \mid m$ (see Theorem 6.3.3., part (b)). Thus we have $m = pk$ and $pn^2 = (pk)^2$ or $pn^2 = p^2k^2$. This gives us $n^2 = pk^2$ which implies $p \mid n^2$. Since p is prime, $p \mid n$. Hence, $(m,n) \neq 1$. This is a contradiction arrived at by assuming $\sqrt{p}$ is a rational number. Therefore, $\sqrt{p}$ is irrational.

■

Theorem 8.3.4

If $a, b \in \mathbf{R}$ and $a < b$, then there exists an irrational number r such that $a < r < b$.

Proof

Let $a, b \in \mathbf{R}$ such that $a < b$ and p a positive prime integer. The $b - a, \sqrt{p} \in \mathbf{R}^+$; and, by Theorem 8.3.1, $\exists n \in \mathbf{N}$ such that $\sqrt{p} < n(b - a)$. Thus we have $\sqrt{p} < nb - na$ or $na + \sqrt{p} < nb$. Let $A = \{k \in \mathbf{N} \mid na < k\sqrt{p}\}$. If $na \leq 0$, then $1 \in A$. If $0 < na$, then $\exists k \in \mathbf{N}$ such that $na < k\sqrt{p}$ and $k \in A$. Hence, $A \neq \emptyset$. Since $\mathbf{N}$ is a well-ordered set and $A \subseteq \mathbf{N}$, A has a smallest element m. Now m is the smallest positive integer such that $na < m\sqrt{p}$ which implies that $(m - 1)\sqrt{p} \leq na$ or $m\sqrt{p} \leq na + \sqrt{p}$. We therefore have $na < m\sqrt{p} \leq na + \sqrt{p} < nb$ or $na < m\sqrt{p} < nb$ which yields $a < \frac{m}{n}\sqrt{p} < b$. Let $r = \frac{m}{n}\sqrt{p}$. Then r is an irrational number (see Exercises 8.3, problem (2)) such that $a < r < b$.

■

Theorems 8.3.2 and 8.3.4 show that the set of rational numbers and the set of irrational numbers are both *dense* in **R**. By this we mean that between any two real numbers there is a rational number and, likewise, between any two real numbers there is an irrational number. Furthermore, Theorem 8.3.4 shows that if we were to represent the rational numbers geometrically as points on a line, there would be a "gap" or "break" between any two rational numbers a and b.

Theorem 8.3.5 (Dedekind Property)

Let A, B $\subseteq$ **R** such that A $\neq \emptyset$ and B $\neq \emptyset$.

If a. **R** = A $\cup$ B and
b. for a $\in$ A and b $\in$ B, a < b,

then there exists c $\in$ **R** (which may belong either to A or B) such that if x < c, then x $\in$ A; and, if x > c, then x $\in$ B.

Proof

Let A, B $\subseteq$ **R** such that A $\neq \emptyset$, B $\neq \emptyset$ and conditions (a) and (b) hold. By condition (b), every element of B is an upper bound for the set A. Thus, by the Axiom of Completeness, A has a least upper bound in **R**. Therefore, there exists c $\in$ **R** such that c = lub A. Hence, a $\leq$ c for all a $\in$ A. Since each element of B is an upper bound of A, c $\leq$ b for all b $\in$ B. Let x $\in$ **R**. If x < c, then x $\notin$ B and hence by condition (a) x $\in$ A. If x > c, then x $\notin$ A and hence by condition (a) x $\in$ B.

■

The number c referred to in Theorem 8.3.5 is called a *Dedekind cut.* Since c = lub A, c is unique for the given sets A and B satisfying the conditions of Theorem 8.3.5. Furthermore, since **R** = A $\cup$ B and since if a $\in$ A and b $\in$ B implies a < b, c $\in$ A or c $\in$ B but c $\notin$ A $\cap$ B. As a consequence, we have **R** = A $\cup$ B and A $\cap$ B = $\emptyset$. Thus A and B are intervals of the form A = $(-\infty,c)$ and B = $[c,\infty)$ or A = $(-\infty,c]$ and B = (c, ∞) again indicating there are no "gaps" or "breaks" in the real line. The real line is "connected."

Theorem 8.3.6 (Nested Interval Property)

For each $i \in \mathbf{N}$, let

a. $I_i = [a_i, b_i] = \{x \in \mathbf{R} \mid a_i \leq x \leq b_i\}$ and

b. $I_i \supseteq I_{i+1}$ $(I_1 \supseteq I_2 \supseteq I_3 \supseteq \ldots)$.

Then $\bigcap_{i \in \mathbf{N}} I_i \neq \emptyset$.

Proof

Let conditions (a) and (b) hold. By condition (b), since $I_i \supseteq I_{i+1}$, we have $a_1 \leq a_2 \leq a_3 \leq \ldots$ and $b_1 \geq b_2 \geq b_3 \geq \ldots$. Let $A = \{a_1, a_2, a_3, \ldots\}$ and $B = \{b_1, b_2, b_3, \ldots\}$. Now each $b_i \in B$ is an upper bound of A; i.e., for a fixed n, $a_i < b_n$ for all $i \in \mathbf{N}$. To see this, suppose $i \leq n$. Then $a_i \leq a_n < b_n$. If $i \geq n$, $a_i < b_i \leq b_n$. Thus, by the Axiom of Completeness, A has a least upper bound in $\mathbf{R}$. Therefore, there exists $c \in \mathbf{R}$ such that $c = \text{lub } A$. As a consequence, we have $a_i \leq c$ for all $i \in \mathbf{N}$; and, since each b_i is an upper bound of A, $c \leq b_i$ for all $i \in \mathbf{N}$.

Thus, $a_i \leq c \leq b_i$ for all $i \in \mathbf{N}$. Hence $c \in I_i$ for all $i \in \mathbf{N}$ and $\bigcap_{i \in \mathbf{N}} I_i \neq \emptyset$.

■

If the length of the intervals I_i in Theorem 8.3.6 approaches 0 as i increases;

i.e., $\lim_{i \to \infty} (b_i - a_i) = 0$, it can be shown that $\bigcap_{i \in \mathbf{N}} I_i = \{c\}$.

Example 8.3.1

a. For $i \in \mathbf{N}$, let $I_i = [0, 1 + 1/i]$. Then $I_1 \supseteq I_2 \supseteq I_3 \supseteq \ldots$ and $\bigcap_{i \in \mathbf{N}} I_i \neq \emptyset$.

In fact $\bigcap_{i \in \mathbf{N}} I_i = [0,1]$.

Example 8.3.1 continued next page.

Example 8.3.1 Continued

b. In Theorem 8.3.6, if the intervals I_i are not closed, then $\bigcap_{i \in \mathbf{N}} I_i$ may be empty.

For $i \in \mathbf{N}$, let $I_i = (1, 1 + 1/i]$. Then $I_1 \supseteq I_2 \supseteq I_3 \supseteq \ldots$ but $\bigcap_{i \in \mathbf{N}} I_i = \emptyset$.

Suppose $\bigcap_{i \in \mathbf{N}} I_i \neq \emptyset$. Then there exists $x \in \bigcap_{i \in \mathbf{N}} I_i$. Hence, $x \in I_i$ for all $i \in \mathbf{N}$. Now $x \in I_i$ implies $x > 1$ or $x - 1 > 0$. By Exercises 8.3, problem (1), there exists $n \in \mathbf{N}$ such that $\frac{1}{n} < x - 1$ or $1 + \frac{1}{n} < x$. Thus, $x \notin I_n$. As a consequence, we have $x \in I_n$ and $x \notin I_n$ which is a contradiction. Therefore, $\bigcap_{i \in \mathbf{N}} I_i = \emptyset$.

c. In Theorem 8.3.6, if the intervals I_i are not of finite length, then I_i may be empty. See Exercises 8.3, problem (5).

Exercises 8.3

1. Show $\forall\ \delta > 0\ \exists\ n \in \mathbf{N}$ such that $\frac{1}{n} < \delta$. (Hint: Use Archimedean Property.)

2. If $m, n, p \in \mathbf{N}$ and p is prime, show $\frac{m}{n}\sqrt{p}$ is an irrational number. (Hint: Assume $\frac{m}{n}\sqrt{p}$ is rational and show $\sqrt{p}$ is rational.)

3. Show the set of rational numbers $\mathbf{Q}$ is not complete. (Hint: Let $a, b \in \mathbf{Q}$ such that $a < b$. Use Theorem 8.3.4 to construct a set $A \subseteq \mathbf{Q}$ which has an upper bound in $\mathbf{Q}$ but not a lub in $\mathbf{Q}$.)

4. For $i \in N$, let $I_i = [1, 1 + 1/i]$. Then

   $I_i \supseteq I_i + 1$ and hence by Theorem 8.3.6,

   $\bigcap_{i \in \mathbf{N}} I_i \neq \emptyset$. Show $\bigcap_{i \in \mathbf{N}} I_i = \{1\}$.

   (Hint: Assume $\bigcap_{i \in \mathbf{N}} I_i \neq \{1\}$.)

5. For $i \in \mathbf{N}$, let $I_i = [i,\infty)$. Show

   $I_i \supseteq I_{i+1}\ \forall i \in \mathbf{N}$, but $\bigcap_{i \in \mathbf{N}} I_i = \emptyset$.

   (Hint: Assume $\bigcap_{i \in \mathbf{N}} I_i \neq \emptyset$.)

Solutions to Odd Exercises

An outline of some of the more important steps are provided. A complete solution may require additional details, an explanation of the steps and/or a unifying conclusion.

Chapter 1
Exercises 1.1

1.

p	$\sim p$	$p \Rightarrow (\sim p)$
T	F	F
F	T	T

3.

p	$\sim p$	$p \wedge \sim p$
T	F	F
F	T	F

5.

p	$\sim p$	$p \vee \sim p$	$\sim(p \vee \sim p)$
T	F	T	F
F	T	T	F

7.

p	q	$p \vee q$	$p \Rightarrow (p \vee q)$
T	T	T	T
T	F	T	T
F	T	T	T
F	F	F	T

9. a. $p \Rightarrow q$
   b. $q \vee r$
   c. $r \Leftrightarrow p$
   d. $q \wedge (\sim r)$

Exercises 1.2

1.

p	q	$p \wedge q$	$(p \wedge q) \Rightarrow p$
T	T	T	T
T	F	F	T
F	T	F	T
F	F	F	T

3.

p	q	$p \wedge q$	$q \Rightarrow (p \wedge q)$	$p \Rightarrow [q \Rightarrow (p \wedge q)]$
T	T	T	T	T
T	F	F	T	T
F	T	F	F	T
F	F	F	T	T

5.

p	q	r	$p \wedge q$	$(p \wedge q) \Rightarrow r$	$q \Rightarrow r$	$p \Rightarrow (q \Rightarrow r)$
T	T	T	T	T	T	T
T	T	F	T	F	F	F
T	F	T	F	T	T	T
T	F	F	F	T	T	T
F	T	T	F	T	T	T
F	T	F	F	T	F	T
F	F	T	F	T	T	T
F	F	F	F	T	T	T

7.

p	q	r	$q \vee r$	$p \Rightarrow (q \vee r)$	$\sim q$	$p \wedge \sim q$	$(p \wedge \sim q) \Rightarrow r$
T	T	T	T	T	F	F	T
T	T	F	T	T	F	F	T
T	F	T	T	T	T	T	T
T	F	F	F	F	T	T	F
F	T	T	T	T	F	F	T
F	T	F	T	T	F	F	T
F	F	T	T	T	T	F	T
F	F	F	F	T	T	F	T

(columns continue below)

$\sim r$	$p \wedge \sim r$	$(p \wedge \sim r) \Rightarrow q$	$[(p \wedge \sim q) \Rightarrow r] \vee [(p \wedge \sim r) \Rightarrow q]$
F	F	T	T
T	T	T	T
F	F	T	T
T	T	F	F
F	F	T	T
T	F	T	T
F	F	T	T
T	F	T	T

Since in each row of the truth table the same truth value appears for the statements being compared, the statements are logically equivalent.

9.

p	q	r	$q \wedge r$	$p \Rightarrow (q \wedge r)$	$p \Rightarrow q$	$p \Rightarrow r$	$(p \Rightarrow q) \wedge (p \Rightarrow r)$
T	T	T	T	T	T	T	T
T	T	F	F	F	T	F	F
T	F	T	F	F	F	T	F
T	F	F	F	F	F	F	F
F	T	T	T	T	T	T	T
F	T	F	F	T	T	T	T
F	F	T	F	T	T	T	T
F	F	F	F	T	T	T	T

11. Converse: $q \Rightarrow p$

    If John passed all his courses, then John studied diligently.

    Inverse: $\sim p \Rightarrow \sim q$

    If John did not study diligently, then John did not pass all his courses.

    Contrapositive: $\sim q \Rightarrow \sim p$

    If John did not pass all his courses, then John did not study diligently.

Exercises 1.3

1. False since for $x = 1$, $|1| = 1 \not> 1$.

   $\sim[(\forall x)(|x| > x)] \equiv (\exists x)\,[\sim(|x| > x)] \equiv (\exists x)\,(|x| \not> x) \equiv (\exists x)\,(|x| \leq x)$

3. False since for $x = 2$, $2^2 - 3(2) = -2 \not> 0$.

   $\sim[(\forall x)(x^2 - 3x > 0)] \equiv (\exists x)\,[\sim(x^2 - 3x > 0)] \equiv (\exists x)(x^2 - 3x \not> 0) \equiv (\exists x)(x^2 - 3x \leq 0)$

5. For $x = 0$, $0^2 - 1 = -1 \not> 0$.

7. For $n = 17$, $17^2 + 17 + 17 = 323 = 17(19)$ and hence is not prime.

9. For $n = 101$, $101 \geq 100$ and n is prime.

Exercises 1.4

1. Direct Method
    1. Hypothesis
    2. Hypothesis
    3. Simplification Law: 2
    4. Detachment Law: 1, 3
    5. Simplification Law: 2
    6. Detachment Law: 4, 5

3. Indirect Method

1.	$p \vee q$	Hypothesis
2.	$\sim q \vee r$	Hypothesis
3.	$\sim(p \vee r)$	Hypothesis
4.	$\sim p \wedge \sim r$	DeMorgan's Law: 3
5.	$\sim p$	Simplification Law: 4
6.	$\sim r$	Simplification Law: 4
7.	q	Detachment Law: 1, 5
8.	$q \wedge \sim r$	Conjunction: 6, 8
9.	$\sim(\sim q \vee r)$	DeMorgan's Law: 8
10.	$(\sim q \vee r) \wedge \sim(\sim q \vee r)$	Contradiction: 2, 9

$\therefore p \vee r$

5. Direct Method

1.	q	Hypothesis
2.	$p \Rightarrow \sim q$	Hypothesis
3.	$r \Rightarrow p$	Hypothesis
4.	$r \Rightarrow \sim q$	Transitive Law: 2, 3
5.	$\sim r$	Detachment: 1, 4

Exercises 1.5

1. Let p(n): $1^2 + 2^2 + \ldots + n^2 = \frac{n(n+1)(2n+1)}{6}$

p(1): $1^2 = 1$ and $\frac{1(1+1)(2 \cdot 1 + 1)}{6} = \frac{1 \cdot 2 \cdot 3}{6} = 1$

$\therefore$ p(1) is true.

p(k): Assume p(k) is true; i.e., $1^2 + 2^2 + \ldots + k^2 = \frac{k(k+1)(2k+1)}{6}$

p(k + 1): $1^2 + 2^2 + \ldots + k^2 + (k+1)^2 = \frac{k(k+1)(2k+1)}{6} + (k+1)^2$

$$= \frac{k(k+1)(2k+1)}{6} + \frac{6(k+1)^2}{6}$$

$$= \frac{(k+1)[k(2k+1) + 6(k+1)]}{6}$$

$$= \frac{(k+1)(2k^2 + 7k + 6)}{6}$$

$$= \frac{(k+1)(k+2)(2k+3)}{6}$$

$$= \frac{(k+1)[(k+1)+1]\,[2(k+1)+1]}{6}$$

$\therefore$ p(k + 1) is true.

Thus, $(\forall n)(p(n))$.

3. Let p(n): $1 \cdot 2 + 2 \cdot 3 + \ldots + n(n+1) = \frac{n(n+1)(n+2)}{3}$

p(1): $1 \cdot 2 = 2$ and $\frac{1(1+1)(1+2)}{3} = \frac{2 \cdot 3}{3} = 2$

$\therefore$ p(1) is true.

p(k): Assume p(k) is true; i.e., $1 \cdot 2 + 2 \cdot 3 + \ldots + k(k+1) = \frac{k(k+1)(k+2)}{3}$

p(k + 1): $1 \cdot 2 + 2 \cdot 3 + \ldots + k(k+1) + (k+1)(k+2) = \frac{k(k+1)(k+2)}{3} + (k+1)(k+2)$

$$= \frac{k(k+1)(k+2)}{3} + \frac{3(k+1)(k+2)}{3}$$

$$= \frac{(k+1)(k+2)(k+3)}{3}$$

$$= \frac{(k+1)[(k+1)+1][(k+1)+2]}{3}$$

$\therefore$ p(k) is true.

Thus, $(\forall n)(p(n))$.

5. Using Mathematical Induction II, for $n \geq 2$ and $a > 0$, let

p(n): $(a+1)^n > na + 1$.

p(2): $(a+1)^2 = a^2 + 2a + 1 > 2a + 1$ since $a^2 > 0$.

$\therefore$ p(2) is true.

p(k): Assume p(k) is true for all k, $2 \leq k < m$; i.e., $(a+1)^k > ka + 1$.

p(m): Since $2 \leq m - 1 < m$, p(m – 1) is true.

$(a+1)^{m-1} > (m-1)a + 1$

$(a+1)^{m-1}(a+1) > [(m-1)a + 1](a+1)$ since $a + 1 > 0$

$(a+1)^m > (ma - a + 1)(a+1)$

$(a+1)^m > ma^2 + ma - a^2 - a + a + 1$

$(a+1)^m > ma^2 - a^2 + ma + 1$

$(a+1)^m > (m-1)a^2 + ma + 1$

But $(m-1)a^2 + ma + 1 > ma + 1$ since $(m-1)a^2 > 0$.

Hence, $(a+1)^m > ma + 1$ by transitivity.

$\therefore$ p(m) is true.

Thus, $(\forall n \geq 2)(p(n))$.

7. Let p(n): n (a + b) = na + nb.

p(1): 1(a + b) = a + b = 1a + 1b.

$\therefore$ p(1) is true.

p(k): Assume p(k) is true; i.e., k (a + b) = ka + kb.

p(k + 1):	(k + 1)(a + b) = k(a + b) + (a + b)	definition
	= ka + kb + a + b	since p(k) is true
	= ka + a + kb + b	commutative property
	= (k + 1)a + (k + 1)b	definition

$\therefore$ p(k + 1) is true.

Thus, $(\forall n)(p(n))$.

9. Let m be an arbitrary fixed natural number and for $n \geq 1$, let

p(n): (mn)a = m(na).

p(1): (m1) a = ma = m(1a)

$\therefore$ p(1) is true.

p(k): Assume p(k) is true; i.e., (mk)a = m(ka).

p(k + 1):	[m(k + 1)]a = (mk + m)a	by problem (7)
	= (mk) a + ma	by problem (8)
	= m(ka) + ma	since p(k) true
	= m (ka + a)	by problem (7)
	= m[(k + 1) a]	definition

$\therefore$ p(k + 1) is true.

Thus, $(\forall n)(p(n))$. Since m is an arbitrary natural number, (mn) a = m(na) is true for all natural numbers m and n.

Chapter 2
Exercises 2.1

1. {6,7,8,9,10,11,12,13}
3. $\emptyset$
5. {. . . , -9,-6,-3,0,3,6,9, . . .}
7. {1,3,5,7, . . .}

Exercises 2.2

1. a. A = {±6,±7,±8, . . .}
   b. B = {2,9,28,65, . . .}
   c. C = {-6,-5,-4,-3,-2,-1,0,1,2}
   d. D = {1,4,9,16,25,36,49,64,81,100,121,144}

3. Hint: Use the given Venn Diagrams to construct your counterexamples.

a.

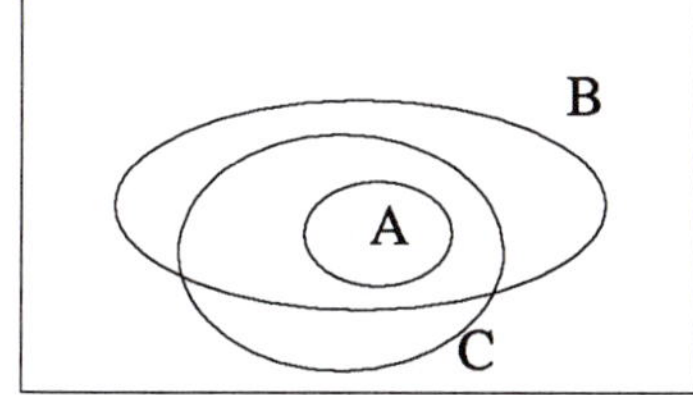

b.

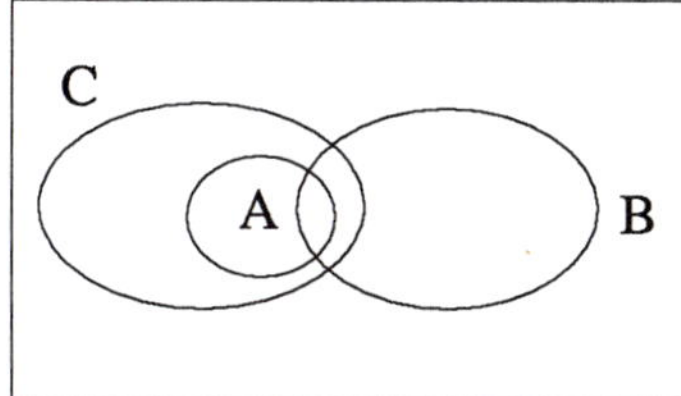

c.

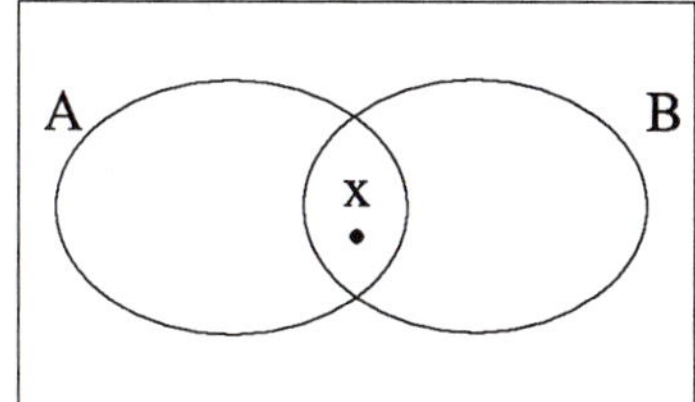

Exercises 2.3

1. $A \subseteq B \Leftrightarrow A \cup B = B$

   ($\Rightarrow$) Let $A \subseteq B$. Show $A \cup B = B$

   Since $B \subseteq A \cup B$, show $A \cup B \subseteq B$.

   Let $x \in A \cup B$. Then $x \in A$ or $x \in B$.

   If $x \in A$, $x \in B$ since $A \subseteq B$.

   If $x \in B$, $x \in B$.

   In either case, $x \in B$. Thus $A \cup B \subseteq B$.

   $\therefore A \cup B = B$.

   ($\Leftarrow$) Let $A \cup B = B$. Show $A \subseteq B$.

   Let $x \in A$. Then $x \in A \cup B = B$.

   Hence, $x \in B$.

   $\therefore A \subseteq B$.

3. a. Let $A \subseteq B$.

$x \in A \cup C \Rightarrow x \in A \vee x \in C$
$\Rightarrow x \in B \vee x \in C$
$\Rightarrow x \in B \cup C$
$\therefore A \cup C \subseteq B \cup C$.

b. Let $A \subseteq B$.

$x \in A \cap C \Rightarrow x \in A \wedge x \in C$
$\Rightarrow x \in B \wedge x \in C$
$\Rightarrow x \in B \cap C$
$\therefore A \cap C \subseteq B \cap C$.

c. Let $A \subseteq B$ and $C \subseteq D$.

$x \in A \cup C \Rightarrow x \in A \vee x \in C$
If $x \in A$, $x \in B$. Thus $x \in B \cup D$.
If $x \in C$, $x \in D$. Thus $x \in B \cup D$.
In either case, $x \in B \cup D$.
$\therefore A \cup C \subseteq B \cup D$.

d. Let $A \subseteq B$ and $C \subseteq D$.

$x \in A \cap C \Rightarrow x \in A \wedge x \in C$
$x \in A \Rightarrow x \in B$
$x \in C \Rightarrow x \in D$
$x \in B \wedge x \in D \Rightarrow x \in B \cap D$
$\therefore A \cap C \subseteq B \cap D$.

5. Let $A' \subseteq B$.

Show $A \cup B \subseteq X$ and $X \subseteq A \cup B$.
Since $A, B \subseteq X$, $A \cup B \subseteq X$.
Let $x \in X$. $X = A \cup A'$. Hence $x \in A \vee x \in A'$.
If $x \in A$, then $x \in A \cup B$ and $X \subseteq A \cup B$.
If $x \in A'$, then $x \in B$ and $x \in A \cup B$. Thus $X \subseteq A \cup B$.
In either case, $X \subseteq A \cup B$.
$\therefore A \cup B = X$.

7. ($\Rightarrow$) Let $A \subseteq B$ and $x \in B'$.

Now $x \in B' \Rightarrow x \notin B \Rightarrow x \notin A \Rightarrow x \notin A'$.
$\therefore B' \subseteq A'$.

($\Leftarrow$) Let $B' \subseteq A'$ and $x \in A$.

Now $x \in A \Rightarrow x \notin A' \Rightarrow x \notin B' \Rightarrow x \in B$.
$\therefore A \subseteq B$.

9. a. $(A - B) \cap (A - C) = (A \cap B') \cap (A \cap C')$
$= (A \cap A) \cap (B' \cap C')$
$= A \cap (B \cup C)'$
$= A - (B \cup C)$

b. $A - (B \cap C) = A \cap (B \cap C)'$
$= A \cap (B' \cup C')$
$= (A \cap B') \cup (A \cap C')$
$= (A - B) \cup (A - C)$

Exercises 2.4

1. Let $Y \in \mathcal{A}$ and $x \in Y$.

Now $x \in \bigcup_{B \in \mathcal{A}} B = \{x \in X \mid \exists B \in \mathcal{A}, x \in B\}$.

Since $Y \in \mathcal{A}$ and $x \in Y$, $x \in \bigcup_{B \in \mathcal{A}} B$.

Thus, $Y \subseteq \bigcup_{B \in \mathcal{A}} B$.

3. a. Let $x \in \bigcup_{B \in \mathcal{A}} B$. Then $\exists B \in \mathcal{A}$ such that $x \in B$. Since $B \subseteq X$, $x \in X$.

Thus, $\bigcup_{B \in \mathcal{A}} B \subseteq X$.

b. Let $x \in \bigcap_{B \in \mathcal{A}} B$. Then $x \in B$, $\forall B \in \mathcal{A}$. Since $B \subseteq X$, $\forall B \in \mathcal{A}$, $x \in X$.

Thus, $\bigcap_{B \in \mathcal{A}} B \subseteq X$.

5. Show $Y \cap \left(\bigcup_{B \in \mathcal{A}} B\right) \subseteq \bigcup_{B \in \mathcal{A}} (Y \cap B)$ and $\bigcup_{B \in \mathcal{A}} (Y \cap B) \subseteq Y \cap \left(\bigcup_{B \in \mathcal{A}} B\right)$.

($\Rightarrow$) Let $x \in Y \cap \left(\bigcup_{B \in \mathcal{A}} B\right) \Rightarrow x \in Y \wedge x \in \bigcup_{B \in \mathcal{A}} B$
$\Rightarrow x \in Y \wedge [\exists B \in \mathcal{A}, x \in B]$
$\Rightarrow \exists B \in \mathcal{A}, x \in Y \wedge x \in B$
$\Rightarrow \exists B \in \mathcal{A}, x \in Y \cap B$
$\Rightarrow x \in \bigcup_{B \in \mathcal{A}} (Y \cap B)$.

(Problem (5) continued next page.)

(Problem (5) continued)

$(\Leftarrow)$ Let $x \in \bigcup_{B \in \mathcal{A}} (Y \cap B) \Rightarrow \exists B \in \mathcal{A}, x \in Y \cap B$

$$\Rightarrow \exists B \in \mathcal{A}, x \in Y \wedge x \in B$$
$$\Rightarrow x \in Y \wedge (\exists B \in \mathcal{A}, x \in B)$$
$$\Rightarrow x \in Y \wedge x \in \bigcup_{B \in \mathcal{A}} B$$
$$\Rightarrow x \in Y \cap \left(\bigcup_{B \in \mathcal{A}} B\right).$$

Exercises 2.5

1. $A \times (B \cup C) = (A \times B) \cup (A \times C)$

$$\begin{aligned}(x,y) \in A \times (B \cup C) &\Leftrightarrow x \in A \wedge y \in B \cup C \\ &\Leftrightarrow x \in A \wedge (y \in B \vee y \in C) \\ &\Leftrightarrow (x \in A \wedge y \in B) \vee (x \in A \wedge y \in C) \\ &\Leftrightarrow (x,y) \in A \times B \vee (x,y) \in A \times C \\ &\Leftrightarrow (x,y) \in (A \times B) \cup (A \times C)\end{aligned}$$

3. $A \times B = \{(1,4),(1,5),(1,6),(2,4),(2,5),(2,6)\}$

$C \times D = \{(1,3),(1,4),(1,5),(1,6),(2,3),(2,4),(2,5),(2,6),(3,3),(3,4),(3,5),(3,6)\}$

$\therefore A \times B \subseteq C \times D$

5. a. $H \times K \times L = \{(p,r,t),(p,r,u),(p,r,v),(p,s,t),(p,s,u),(p,s,v),(q,r,t),(q,r,u)(q,r,v),(q,s,t),(q,s,u),(q,s,v)\}$

b. $H \times (K \times L) = \{p,q\} \times \{(r,t),(r,u),(r,v),(s,t),(s,u),(s,v)\}$

$= \{(p,(r,t)),(p,(r,u)),(p,(r,v)),(p,(s,t)),(p,(s,u)),(p,(s,v)),(q,(r,t)),$
$(q,(r,u)),(q,(r,v)),(q,(s,t)),(q,(s,u)),(q,(s,v))\}$

c. $(H \times K) \times L = \{(p,r),(p,s),(q,r),(q,s)\} \times \{t,u,v\}$

$= \{((p,r),t),((p,r),u),((p,r),v),((p,s),t),((p,s),u),((p,s),v),((q,r),t),$
$((q,r),u),((q,r),v),((q,s)t),((q,s,u),((q,s),v)\}$

d. The sets are not the same since $(p,r,t) \neq (p,(r,t)) \neq ((p,r),t)$, but all the sets have the same number of elements.

Chapter 3
Exercises 3.1

1. Each element of A can be assigned to either a or b; hence, there are $2^3 = 8$ functions from A to B.

$f_1 : A \longrightarrow B$: $1 \to a$, $2 \to a$, $3 \to a$

$f_2 : A \longrightarrow B$: $1 \to b$, $2 \to b$, $3 \to b$

$f_3 : A \longrightarrow B$: $1 \to a$, $2 \to b$, $3 \to b$

$f_4 : A \longrightarrow B$: $1 \to b$, $2 \to a$, $3 \to b$

$f_5 : A \longrightarrow B$: $1 \to b$, $2 \to b$, $3 \to a$

$f_6 : A \longrightarrow B$: $1 \to b$, $2 \to a$, $3 \to a$

$f_7 : A \longrightarrow B$: $1 \to a$, $2 \to b$, $3 \to a$

$f_8 : A \longrightarrow B$: $1 \to a$, $2 \to a$, $3 \to b$

Exercises 3.2

1. a. $f_3, f_4, f_5, f_6, f_7, f_8$
   b. No. A has more elements than B.
   c. n^m

3.

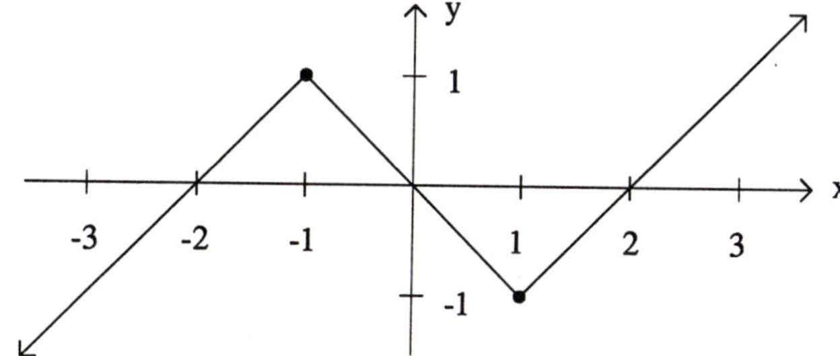

Visually, we can see that f is onto since each horizontal line will intersect the graph at least once.

Alternately, by proof, let $y \in \mathbf{R}$. Show $\exists\, x \in \mathbf{R}$ such that $f(x) = y$.

For $y \leq -1$, let $x = y - 2 \leq -3$ and $f(x) = y$.

For $-1 < y < 1$, let $x = -y$ and $f(x) = y$.

For $y \geq 1$, let $x = y + 2 \geq 3$ and $f(x) = y$.

Since $R_f = \mathbf{R}$, f is onto.

f is not 1–1 since $-2, 2 \in D_f = \mathbf{R}$ and $-2 \neq 2$, but $f(-2) = f(2) = 0$.

5. f: $A \to A \times \{b\}$ where $f(a) = (a,b)$.
f is 1–1:
Let $a_1, a_2 \in A$ such that $f(a_1) = f(a_2)$.
Then $(a_1,b) = (a_2,b)$ and hence, $a_1 = a_2$.

f is onto:
Let $(x,b) \in A \times \{b\}$.
Then $x \in A$ and $f(x) = (x,b)$.

7. A generalization of problem (6) yields: The function
$f: (A_1 x \ldots x A_{n-1}) \times A_n \to A_1 x \ldots x A_{n-1} \times A_n$ where $f((a_1, \ldots, a_{n-1}), a_n) = (a_1, \ldots, a_{n-1}, a_n)$ is a 1–1 correspondence. Hence, $(A_1 \times \ldots \times A_{n-1}) \times A_n$ and $A_1 \times \ldots \times A_{n-1} \times A_n$ have the same number of elements.

For $n \geq 1$, let p(n): If $B = \{0,1\}$, then B^n has 2^n elements.
p(1): If $B = \{0,1\}$, then $B^1 = B = \{0,1\}$ has $2 = 2^1$ elements.
∴ p(1) is true.

p(k): Assume p(k) is true; i.e., if $B = \{0,1\}$, then B^k has 2^k elements.

p(k + 1): Let $B = \{0,1\}$. Show p(k + 1) is true.
By Theorem 3.2.1, $B^k \times B$ has $2^k \bullet 2 = 2^{k+1}$ elements.
But $B^k \times B = (Bx \ldots xB) \times B$ and $B^{k+1} = Bx \ldots x B \times B$ have the same number of elements. Thus B^{k+1} has 2^{k+1} elements.
∴ p(k + 1) is true.
Thus $(\forall n)(p(n))$ is true.

9. Observe the results of problem (8), part (c).
f is 1–1:
Let $A = \{a_1, a_2, \ldots, a_n\}$ and $X, Y \in P(A)$ such that $f(X) = f(Y)$.
Then $(b_1, \ldots, b_n) = (b_1', \ldots, b_n')$ and $b_i = b_i'$ for $i = 1, \ldots, n$.
We must show X = Y. Let $x \in X$. Since $X \subseteq A$, $x = a_k$ for some k, $1 \leq k \leq n$. Hence $b_k = 1 = b_k'$, $x = a_k \in Y$ and $X \subseteq Y$. Let $y \in Y$. Since $Y \subseteq A$, $y = a_j$ for some j, $1 \leq j \leq n$. Hence $b_j' = 1 = b_j$, $y = a_j \in X$ and $Y \subseteq X$.
Therefore, X = Y and f is 1–1.

f is onto:
Let $(b_1, \ldots, b_n) \in B^n$.
If $b_i = 0$ for each $i = 1, \ldots, n$, then $X = \emptyset \in P(A)$ and $f(\emptyset) = (0, \ldots, 0)$.
If not all $b_i = 0$, then $b_i = 1$ for some i. Construct the set X as follows:
Let $a_i \in X \Leftrightarrow b_i = 1$. Then $X \in P(A)$ and $f(X) = (b_1, \ldots, b_n)$.

Hence, f is 1–1 and onto and therefore P(A) has 2^n elements.

Exercises 3.3

1. Each f_i must be defined for all reals; and if $n \in \mathbf{N}$, then $f_i(n)$ must be n^2.

$f_1(x) = x^2 \qquad f_2(x) = \begin{cases} x^2 & \text{if } x \in \mathbf{N} \\ x & \text{if } x \in \mathbf{R}-\mathbf{N} \end{cases} \qquad f_3(x) = \begin{cases} x^2 & \text{if } x \in \mathbf{N} \\ 0 & \text{if } x \in \mathbf{R}-\mathbf{N} \end{cases}$

(Note: There are many other choices for f_1, f_2, f_3.)

3. $D_f \cap D_g = (-\infty, o] \cap \mathbf{R} = (-\infty, 0]$
$f(x) = g(x)\ \forall\ x \in D_f \cap D_g$

$(f \cup g)\ (x) = g(x) = |x|$

5. a. $D_{f \circ \mathbf{e}_A} = D_{\mathbf{e}_A} = A = D_f$
$(f \circ \mathbf{e}_A)(a) = f(\mathbf{e}_A(a)) = f(a)\ \forall\ a \in A$
$\therefore\ f \circ \mathbf{e}_A = f$

b. $D_{\mathbf{e}_B \circ f} = D_f = A$
$(\mathbf{e}_B \circ f)(a) = \mathbf{e}_B(f(a)) = f(a)\ \forall\ a \in A$
since $f(a) \in B$
$\therefore\ \mathbf{e}_B \circ f = f$

Exercises 3.4

1. $D_{f \circ f^{-1}} = D_{f^{-1}} = R_f = B = D_{\mathbf{e}_B}$
Let $b \in B$. Since f is onto, $\exists\ a \in A$ such that $f(a) = b$.
Since f^{-1} exists, $a = f^{-1}(b)$.
Now $(f \circ f^{-1})(b) = f(f^{-1}(b)) = f(a) = b = \mathbf{e}_B(b)$
$\therefore\ f \circ f^{-1} = \mathbf{e}_B$

3. $f^{-1}: \mathbf{R} \to \mathbf{R}$. Let $x \in D_{f^{-1}} = \mathbf{R}$. Then $f^{-1}(x) \in R_{f^{-1}} = \mathbf{R}$.
Let $f^{-1}(x) = y$. Then $x = f(y) = 3y + 5$ or $y = \frac{x-5}{3}$.
Hence, $f^{-1}(x) = \frac{x-5}{3}$.

5. $f^{-1}(x) = \frac{x-5}{3}$ and $g^{-1}(x) = \sqrt[3]{x}$.
$(f \circ g)(x) = f(g(x)) = 3x^3 + 5$
$(f \circ g)^{-1}(x) = \sqrt[3]{\frac{x-5}{3}}$
$(g^{-1} \circ f^{-1})(x) = g^{-1}(f^{-1}(x)) = g^{-1}\left(\frac{x-5}{3}\right) = \sqrt[3]{\frac{x-5}{3}}$
Thus, $(f \circ g)^{-1} = g^{-1} \circ f^{-1}$.

Exercises 3.5

1. a. $f(X) = \{3,4,6\}$
   $f^{-1}(f(X)) = \{a,c,d,e\}$

   b. $f^{-1}(Y) = \{a,b,c\}$
   $f(f^{-1}(Y)) = \{1,4\}$

3. Let $f: A \to B$ and $Y \subseteq B$.

   ($\Rightarrow$) Show $f(f^{-1}(Y)) \subseteq f(A) \cap Y$
   Let $b \in f(f^{-1}(Y))$. Then $b = f(a)$ for some $a \in f^{-1}(Y)$.
   But $a \in f^{-1}(Y) \Rightarrow f(a) \in Y \Rightarrow b \in Y$. Also, $a \in f^{-1}(Y) \subseteq A \Rightarrow f(a) \in f(A) \Rightarrow b \in f(A)$.
   Hence, $b \in Y \cap f(A)$ and $f(f^{-1}(Y)) \subseteq Y \cap f(A)$.

   ($\Leftarrow$) Show $f(A) \cap Y \subseteq f(f^{-1}(Y))$.
   Let $b \in f(A) \cap Y$. Then $b \in f(A)$ and $b \in Y$.
   $b \in f(A) \Rightarrow b = f(a)$ for some $a \in A$.
   $b = f(a) \in Y \Rightarrow a \in f^{-1}(Y) \Rightarrow b = f(a) \in f(f^{-1}(Y))$.
   Hence, $f(A) \cap Y \subseteq f(f^{-1}(Y))$.

   $\therefore f(f^{-1}(Y)) = f(A) \cap Y$.

5. By Theorem 3.5.3, part (b), $f(X \cap Z) \subseteq f(X) \cap f(Z)$.
   Must show if f is 1–1, $f(X) \cap f(Z) \subseteq f(X \cap Z)$.
   Let $b \in f(X) \cap f(Z)$. Then $b \in f(X)$ and $b \in f(Z)$. $b \in f(X) \Rightarrow b = f(a_1)$ for some $a_1 \in X$.
   $b \in f(Z) \Rightarrow b = f(a_2)$ for some $a_2 \in Z$. Since f is 1–1 and $f(a_1) = b = f(a_2)$, $a_1 = a_2$ and
   $a_1 = a_2 \in X \cap Z$. Hence, $b = f(a_1) = f(a_2) \in f(X \cap Z)$ and $f(X) \cap f(Z) \subseteq f(X \cap Z)$.
   $\therefore f(X \cap Z) = f(X) \cap f(Z)$.

7. Use problem (6), part (a).
   $$\begin{aligned} f^{-1}(Y') &= f^{-1}(B - Y) \\ &= f^{-1}(B) - f^{-1}(Y) \\ &= A - f^{-1}(Y) \\ &= [f^{-1}(Y)]' \end{aligned}$$

Chapter 4
Exercises 4.1

1. A x B = {(a,1),(a,2),(b,1),(b,2)}. There are $2^4 = 16$ relations from A to B.
   Listed are 4 of the 16 relations: Ø; {(a,1)}; {(a,2),(b,1)}; {(a,1),(a,2),(b,2)}.

3. {(a,1),(b,1)}; {(a,1),(b,2)}; {(a,2),(b,1)}; {(a,2),(b,2)}

5. a. $D_f = \{a \in A \mid (a,b) \in f\} = A$
   b. $R_f = \{b \in B \mid (a,b) \in f\} \subseteq B$
   c. f is 1–1 if, whenever $(a_1,b) \in f$ and $(a_2,b) \in f$, then $a_1 = a_2$.
   d. Let f be 1–1 and onto. Then $f^{-1} = \{(b,a) \mid (a,b) \in f\}$ is called the inverse function of f.
   e. If f: A → B and g: B → C, then g ∘ f: A → C is defined as:
      $g \circ f = \{(a,c) \mid \exists\, b \in B$ such that $(a,b) \in f$ and $(b,c) \in g\}$.

Exercises 4.2

1. A x A = {(1,1),(1,2),(2,1),(2,2)}. There are $2^4 = 16$ relations that can be defined on A.
   Listed are 4 of the 16 relations: Ø; {(2,1)}; {(1,1),(2,2)}; {(1,2),(2,1),(2,2)}.

3. {(1,1),(2,2)}; {(1,2),(2,1)}

Exercises 4.3

1. a. $\mathcal{R}$ is not reflexive since $c \in A$ but $(c,c) \notin \mathcal{R}$. $\mathcal{R}$ is symmetric.
      $\mathcal{R}$ is not transitive since $(c,a) \in \mathcal{R}$ and $(a,c) \in \mathcal{R}$ but $(c,c) \notin \mathcal{R}$.
   b. $\mathcal{R}$ is reflexive since $x = x \cdot 1 \Rightarrow x \,\mathcal{R}\, x$. $\mathcal{R}$ is not symmetric since $2 \,\mathcal{R}\, 6$ but $6 \not\mathcal{R}\, 2$.
      $\mathcal{R}$ is transitive since, if $x \,\mathcal{R}\, y$ and $y \,\mathcal{R}\, z$, then $y = xm$, $z = yn$ and
      $z = yn = (xm)n = x(mn) = xk$ where $k = mn \Rightarrow x \,\mathcal{R}\, z$.

3. $\mathcal{R}$ is reflexive since $x - x = 0 \in \mathbf{Z} \Rightarrow x \,\mathcal{R}\, x$. $\mathcal{R}$ is symmetric since, if $x \,\mathcal{R}\, y$,
   then $x - y \in \mathbf{Z} \Rightarrow -(x - y) = y - x \in \mathbf{Z} \Rightarrow y \,\mathcal{R}\, x$. $\mathcal{R}$ is transitive since, if $x \,\mathcal{R}\, y$
   and $y \,\mathcal{R}\, z$, then $x - y, y - z \in \mathbf{Z} \Rightarrow (x - y) + (y - z) = x - z \in \mathbf{Z} \Rightarrow x \,\mathcal{R}\, z$.

5. $\mathcal{R}$ is reflexive since f(a) = f(a) ⇒ f $\mathcal{R}$ f. $\mathcal{R}$ is symmetric since, if f $\mathcal{R}$ g, then f(a) = g(a) ⇒ g(a) = f(a) ⇒ g $\mathcal{R}$ f. $\mathcal{R}$ is transitive since, if f $\mathcal{R}$ g and g $\mathcal{R}$ h, then f(a) = g(a) and g(a) = h(a) ⇒ f(a) = h(a) ⇒ f $\mathcal{R}$ h.

Exercises 4.4

1. a. $\overline{a} = \{x \in A \mid x \mathcal{R} a\}$. Therefore, $\overline{a} \subseteq A$.
   b. $\mathcal{R}$ an equivalence relation on A ⇒ $\mathcal{R}$ is reflexive ⇒ $\forall a \in A, a \mathcal{R} a \Rightarrow a \in \overline{a}$.
   c. $x, y \in \overline{a} \Rightarrow x \mathcal{R} a$ and $y \mathcal{R} a \Rightarrow x \mathcal{R} a$ and $a \mathcal{R} y \Rightarrow x \mathcal{R} y$.

Exercises 4.5

1. $\{\overline{1},\overline{2},\overline{3},\overline{4}\} = \{\overline{1},\overline{2}\}$ where $\overline{1} = \{1,3\}$ and $\overline{2} = \{2,4\}$.

3. a. Hint: See Example 4.3.2, part (b).
   b. $\{\overline{1},\overline{2},\overline{3},\overline{4},\overline{5},\overline{6}\} = \{\overline{1},\overline{2},\overline{3}\}$ where $\overline{1} = \{1,5\}$ and $\overline{2} = \{2,6\}$ and $\overline{3} = \{3,4\}$.

Chapter 5
Exercises 5.1

1. # is an operation on **R** since $y^2 + 1 \neq 0\ \forall y \in \mathbf{R}$.
3. # is an operation on $\mathbf{R}^+$ since $xe^y \in \mathbf{R}^+\ \forall x, y \in \mathbf{R}^+$.
5. # is not an operation on **Z** since x = 2, y = -4 ∈ **Z**, but $2 \# (-4) = 2^{2+(-4)} = 2^{-2} = \frac{1}{4} \notin \mathbf{Z}$.

Exercises 5.2

1. commutative:
$$\begin{aligned} x * y &= x + y - xy \\ &= y + x - yx \\ &= y * x \end{aligned}$$

associative: $x * (y * z) = x * (y + z - yz)$
$= x + (y + z - yz) - x(y + z - yz)$
$= x + y + z - yz - xy - xz + xyz$

$(x * y) * z = (x + y - xy) * z$
$= (x + y - xy) + z - (x + y - xy)z$
$= x + y - xy + z - xz - yz + xyz$
$= x + y + z - yz - xy - xz + xyz$

$\therefore x * (y * z) = (x * y) * z$

3. Let $x = (a,b)$, $y = (c,d)$ and $z = (e,f) \in \mathbf{R} \times \mathbf{R}$.

a. commutative: $x + y = (a,b) + (c,d)$
$= (a + c, b + d)$
$= (c + a, d + b)$
$= (c,d) + (a,b)$
$= y + x$

associative: $(x + y) + z = [(a,b) + (c,d)] + (e,f)$
$= (a + c, b + d) + (e,f)$
$= ((a + c) + e, (b + d) + f)$
$= (a + (c + e), b + (d + f))$
$= (a,b) + (c + e, d + f)$
$= (a,b) + [(c,d) + (e,f)]$
$= x + (y + z)$

b. commutative: $xy = (a,b)(c,d)$
$= (ac, bd)$
$= (ca, db)$
$= (c,d)(a,b)$
$= yx$

associative: $(xy)z = [(a,b)(c,d)](e,f)$
$= (ac, bd)(e,f)$
$= ((ac)e, (bd)f)$
$= (a(ce), b(df))$
$= (a,b)(ce, df)$
$= (a,b)[(c,d), (e,f)]$
$= x(yz)$

5. Let $x = \begin{bmatrix} a_1 & a_2 \\ a_3 & a_4 \end{bmatrix}$ $y = \begin{bmatrix} b_1 & b_2 \\ b_3 & b_4 \end{bmatrix}$ $z = \begin{bmatrix} c_1 & c_2 \\ c_3 & c_4 \end{bmatrix}$.

a. commutative:

$$x + y = \begin{bmatrix} a_1 & a_2 \\ a_3 & a_4 \end{bmatrix} + \begin{bmatrix} b_1 & b_2 \\ b_3 & b_4 \end{bmatrix}$$

$$= \begin{bmatrix} a_1 + b_1 & a_2 + b_2 \\ a_3 + b_3 & a_4 + b_4 \end{bmatrix}$$

$$= \begin{bmatrix} b_1 + a_1 & b_2 + a_2 \\ b_3 + a_3 & b_4 + a_4 \end{bmatrix}$$

$$= \begin{bmatrix} b_1 & b_2 \\ b_3 & b_4 \end{bmatrix} + \begin{bmatrix} a_1 & a_2 \\ a_3 & a_4 \end{bmatrix}$$

$$= y + x$$

associative:

$$(x + y) + z = \left(\begin{bmatrix} a_1 & a_2 \\ a_3 & a_4 \end{bmatrix} + \begin{bmatrix} b_1 & b_2 \\ b_3 & b_4 \end{bmatrix} \right) + \begin{bmatrix} c_1 & c_2 \\ c_3 & c_4 \end{bmatrix}$$

$$= \begin{bmatrix} a_1 + b_1 & a_2 + b_2 \\ a_3 + b_3 & a_4 + b_4 \end{bmatrix} + \begin{bmatrix} c_1 & c_2 \\ c_3 & c_4 \end{bmatrix}$$

$$= \begin{bmatrix} (a_1 + b_1) + c_1 & (a_2 + b_2) + c_2 \\ (a_3 + b_3) + c_3 & (a_4 + b_4) + c_4 \end{bmatrix}$$

$$= \begin{bmatrix} a_1 + (b_1 + c_1) & a_2 + (b_2 + c_2) \\ a_3 + (b_3 + c_3) & a_4 + (b_4 + c_4) \end{bmatrix}$$

$$= \begin{bmatrix} a_1 & a_2 \\ a_3 & a_4 \end{bmatrix} + \begin{bmatrix} b_1 + c_1 & b_2 + c_2 \\ b_3 + c_3 & b_4 + c_4 \end{bmatrix}$$

$$= \begin{bmatrix} a_1 & a_2 \\ a_3 & a_4 \end{bmatrix} + \left(\begin{bmatrix} b_1 & b_2 \\ b_3 & b_4 \end{bmatrix} + \begin{bmatrix} c_1 & c_2 \\ c_3 & c_4 \end{bmatrix} \right)$$

$$= x + (y + z)$$

b. commutative:

$$yx = \begin{bmatrix} b_1 & b_2 \\ b_3 & b_4 \end{bmatrix} \cdot \begin{bmatrix} a_1 & a_2 \\ a_3 & a_4 \end{bmatrix} = \begin{bmatrix} b_1a_1 + b_2a_3 & b_1a_2 + b_2a_4 \\ b_3a_1 + b_4a_3 & b_3a_2 + b_4a_4 \end{bmatrix}$$

Now $xy \neq yx \ \forall x, y \in M$.

$$xy = \begin{bmatrix} 1 & 0 \\ 2 & -1 \end{bmatrix} \begin{bmatrix} 1 & 0 \\ 1 & 1 \end{bmatrix} = \begin{bmatrix} 1 & 0 \\ 1 & -1 \end{bmatrix}$$

BUT

$$yx = \begin{bmatrix} 1 & 0 \\ 1 & 1 \end{bmatrix} \begin{bmatrix} 1 & 0 \\ 2 & -1 \end{bmatrix} = \begin{bmatrix} 1 & 0 \\ 3 & -1 \end{bmatrix}$$

associative:

$$(xy)z = \begin{bmatrix} (a_1b_1 + a_2b_3)c_1 + (a_1b_2 + a_2b_4)c_3 & (a_1b_1 + a_2b_3)c_2 + (a_1b_2 + a_2b_4)c_4 \\ (a_3b_1 + a_4b_3)c_1 + (a_3b_2 + a_4b_4)c_3 & (a_3b_1 + a_4b_3)c_2 + (a_3b_2 + a_4b_4)c_4 \end{bmatrix}$$

$$x(yz) = \begin{bmatrix} a_1(b_1c_1 + b_2c_3) + a_2(b_3c_1 + b_4c_3) & a_1(b_1c_2 + b_2c_4) + a_2(b_3c_2 + b_4c_4) \\ a_3(b_1c_1 + b_2c_3) + a_4(b_3c_1 + b_4c_3) & a_3(b_1c_2 + b_2c_4) + a_4(b_3c_2 + b_4c_4) \end{bmatrix}$$

Show these are equal.

Exercises 5.3

1. Does $\exists\, e \in \mathbf{R} - \{1\}$ such that $x * e = x = e * x\ \forall x \in \mathbf{R} - \{1\}$?
   $x * e = x \Rightarrow x + e - xe = x \Rightarrow e(1 - x) = 0$. Since $1 \notin \mathbf{R} - \{1\}$, $1 - x \neq 0$. Therefore, $e = 0$. Now $*$ commutative and hence $e = 0$ is the identity element for $R - \{1\}$ relative to $*$.

3. a. $e = (0,0)$
   b. Does $\exists\, e \in \mathbf{R} \times \mathbf{R}$ such that $xe = x = ex\ \forall x \in \mathbf{R} \times \mathbf{R}$? Let $x = (a,b)$ and $e = (r,s)$. $xe = x \Rightarrow (a,b)(r,s) = (a,b) \Rightarrow (ar,bs) = (a,b) \Rightarrow ar = a$ and $bs = b \Rightarrow r = s = 1 \Rightarrow e = (1,1)$. Now $\bullet$ is commutative and hence $e = (1,1)$ is the identity element for $\mathbf{R} \times \mathbf{R}$ relative to $\bullet$.

5. a. Does $\exists\, e \in M$ such that $x + e = x = e + x\ \forall x \in M$?

Let $x = \begin{bmatrix} a_1 & a_2 \\ a_3 & a_4 \end{bmatrix}$ and $e = \begin{bmatrix} e_1 & e_2 \\ e_3 & e_4 \end{bmatrix}$.

$$x + e = x \Rightarrow \begin{bmatrix} a_1 + e_1 & a_2 + e_2 \\ a_3 + e_3 & a_4 + e_4 \end{bmatrix} = \begin{bmatrix} a_1 & a_2 \\ a_3 & a_4 \end{bmatrix}$$

$\Rightarrow a_1 + e_1 = a_1,\ a_2 + e_2 = a_2,\ a_3 + e_3 = a_3,\ a_4 + e_4 = a_4$

$\Rightarrow e_1 = e_2 = e_3 = e_4 = 0$

$$\Rightarrow e = \begin{bmatrix} 0 & 0 \\ 0 & 0 \end{bmatrix}.$$

Now + is commutative and hence $e = \begin{bmatrix} 0 & 0 \\ 0 & 0 \end{bmatrix}$ is the identity element for M relative to +.

b. Does $\exists\, e \in M$ such that $xe = x = ex\ \forall x \in M$?

Let $x = \begin{bmatrix} a_1 & a_2 \\ a_3 & a_4 \end{bmatrix}$ and $e = \begin{bmatrix} e_1 & e_2 \\ e_3 & e_4 \end{bmatrix}$.

$$xe = x \Rightarrow \begin{bmatrix} a_1 e_1 + a_2 e_3 & a_1 e_2 + a_2 e_4 \\ a_3 e_1 + a_4 e_3 & a_3 e_2 + a_4 e_4 \end{bmatrix} = \begin{bmatrix} a_1 & a_2 \\ a_3 & a_4 \end{bmatrix}$$

$\Rightarrow a_1e_1 + a_2e_3 = a_1,\ \ a_1e_2 + a_2e_4 = a_2,\ \ a_3e_1 + a_4e_3 = a_3$ and $a_3e_2 + a_4e_4 = a_4$.

Solving the equations $a_1e_1 + a_2e_3 = a_1$ and $a_3e_1 + a_4e_3 = a_3$ for e_1 and e_2, we get $e_1 = 1, e_3 = 0$. Solving the equations $a_1e_2 + a_2e_4 = a_2$ and $a_3e_2 + a_4e_4 = a_4$ for e_2 and e_4, we get $e_2 = 0$ and $e_4 = 1$.

If $e = \begin{bmatrix} 1 & 0 \\ 0 & 1 \end{bmatrix}$, then $xe = x$.

Since multiplication is not commutative, we must also verify $ex = x$. Hence, e is the identity element for M relative to •.

Exercises 5.4

1. (Problem 1) Let $x \in \mathbf{R} - \{1\}$. Does $\exists\, y \in \mathbf{R} - \{1\}$ such that $x * y = 0 = y * x$?
$x * y = 0 \Rightarrow x + y - xy = 0 \Rightarrow y\,(1 - x) = -x \Rightarrow y = \frac{-x}{1 - x} = \frac{x}{x - 1}$ since $1 - x \neq 0$.
Now $\frac{x}{x - 1} \in \mathbf{R} - \{1\}$; and, since $*$ is commutative, $\frac{x}{x - 1}$ is the inverse of x relative to $*$.

(Problem 3a) Let $x \in \mathbf{R} \times \mathbf{R}$. Does $\exists\, y \in \mathbf{R} \times \mathbf{R}$ such that $x + y = (0,0) = y + x$?
Let $x = (a,b)$ and $y = (r,s)$. $x + y = (0,0) \Rightarrow (a,b) + (r,s) = (0,0) \Rightarrow (a + r, b + s) = (0,0) \Rightarrow$ $a + r = 0$ and $b + s = 0 \Rightarrow r = -a$ and $s = -b$. Now $y = (-a,-b) \in \mathbf{R} \times \mathbf{R}$; and, since $+$ is commutative, $y = (-a,-b)$ is the inverse of $x = (a,b)$ relative to $+$.

(Problem 3b) Let $x \in \mathbf{R} \times \mathbf{R}$. Does $\exists\, y \in \mathbf{R} \times \mathbf{R}$ such that $xy = (1,1) = yx$?
Let $x = (a,b)$ and $y = (r,s)$. $xy = (1,1) \Rightarrow (a,b)(r,s) = (1,1) \Rightarrow (ar,bs) = (1,1) \Rightarrow ar = 1$ and $bs = 1$. If $a \neq 0$ and $b \neq 0$, then $r = \frac{1}{a}$ and $s = \frac{1}{b}$. Hence, if $a \neq 0$ and $b \neq 0$, then $y = (\frac{1}{a},\frac{1}{b}) \in \mathbf{R} \times \mathbf{R}$; and, since $\bullet$ is commutative, $y = (\frac{1}{a},\frac{1}{b})$ is the inverse of $x = (a,b)$ relative to $\bullet$.

(Problem 5a) Let $x \in \mathbf{M}$. Does $\exists\, y \in \mathbf{M}$ such that $x + y = \begin{bmatrix} 0 & 0 \\ 0 & 0 \end{bmatrix} = y + x$?

Let $x = \begin{bmatrix} a_1 & a_2 \\ a_3 & a_4 \end{bmatrix}$ and $y = \begin{bmatrix} y_1 & y_2 \\ y_3 & y_4 \end{bmatrix}$.

$$x + y = \begin{bmatrix} 0 & 0 \\ 0 & 0 \end{bmatrix} \Rightarrow \begin{bmatrix} a_1 + y_1 & a_2 + y_2 \\ a_3 + y_3 & a_4 + y_4 \end{bmatrix} = \begin{bmatrix} 0 & 0 \\ 0 & 0 \end{bmatrix} \Rightarrow$$

$a_1 + y_1 = 0$, $a_2 + y_2 = 0$, $a_3 + y_3 = 0$, $a_4 + y_4 = 0 \Rightarrow y_1 = -a_1$, $y_2 = -a_2$, $y_3 = -a_3$, $y_4 = -a_4$.

Now $y = \begin{bmatrix} -a_1 & -a_2 \\ -a_3 & -a_4 \end{bmatrix} \in$ M; and, since $+$ is commutative, $y = \begin{bmatrix} -a_1 & -a_2 \\ -a_3 & -a_4 \end{bmatrix}$ is the inverse of

$x = \begin{bmatrix} a_1 & a_2 \\ a_3 & a_4 \end{bmatrix}$ relative to $+$.

(Problem 5b) Let $x \in \mathbf{M}$. Does $\exists\, y \in \mathbf{M}$ such that $xy = \begin{bmatrix} 1 & 0 \\ 0 & 1 \end{bmatrix} = yx$?

Let $x = \begin{bmatrix} a_1 & a_2 \\ a_3 & a_4 \end{bmatrix}$ and $y = \begin{bmatrix} y_1 & y_2 \\ y_3 & y_4 \end{bmatrix}$.

$$xy = \begin{bmatrix} 1 & 0 \\ 0 & 1 \end{bmatrix} \Rightarrow \begin{bmatrix} a_1y_1 + a_2y_3 & a_1y_2 + a_2y_4 \\ a_3y_1 + a_4y_3 & a_3y_2 + a_4y_4 \end{bmatrix} = \begin{bmatrix} 1 & 0 \\ 0 & 1 \end{bmatrix}$$

$\Rightarrow a_1y_1 + a_2y_3 = 1,\ \ a_1y_2 + a_2y_4 = 0,\ \ a_3y_1 + a_4y_3 = 0$ and $a_3y_2 + a_4y_4 = 1$.

Solving the equations $a_1y_1 + a_2y_3 = 1$ and $a_3y_1 + a_4y_3 = 0$ for y_1 and y_3; and $a_1y_2 + a_2y_4 = 0$ and $a_3y_2 + a_4y_4 = 1$ for y_2 and y_4, we get
$y_1 = \dfrac{a_4}{a_1a_4 - a_2a_3}$, $y_2 = \dfrac{-a_2}{a_1a_4 - a_2a_3}$, $y_3 = \dfrac{-a_3}{a_1a_4 - a_2a_3}$ and $y_4 = \dfrac{a_1}{a_1a_4 - a_2a_3}$, if $a_1a_4 - a_2a_3 \neq 0$.

If $y = \begin{bmatrix} \dfrac{a_4}{a_1a_4 - a_2a_3} & \dfrac{-a_2}{a_1a_4 - a_2a_3} \\ \dfrac{-a_3}{a_1a_4 - a_2a_3} & \dfrac{a_1}{a_1a_4 - a_2a_3} \end{bmatrix}$, then $xy = e$.

Since multiplication is not commutative, we must also verify $yx = e$.
Hence, y is the inverse of x relative to •.

3. a. yes
   b. yes, b = identity element
   c.

element	inverse
a	d
b	b
c	c
d	a

Chapter 6
Exercises 6.1

1. a. $a, b \in \mathbf{E} \Rightarrow a = 2m, b = 2n \Rightarrow a + b = 2m + 2n = 2(m + n) = 2k$
   where $k = m + n \Rightarrow a + b \in \mathbf{E}$.

   c. $a \in \mathbf{E}$ and $b \in \mathbf{O} \Rightarrow a = 2m, b = 2n + 1 \Rightarrow a + b = 2m + (2n + 1) = 2(m + n) + 1 = 2k + 1$
   where $k = m + n \Rightarrow a + b \in \mathbf{O}$.

3. b. $d \mid a \Rightarrow a = dc \Rightarrow a = (-d)(-c) \Rightarrow -d \mid a$.

   d. $a \mid b$ and $b \mid a \Rightarrow b = ac$ and $a = bd \Rightarrow b = (bd)c = b(dc) \Rightarrow 1 = dc \Rightarrow$
   $d = c = 1$ or $d = c = -1 \Rightarrow a = b$ or $a = -b$.

   f. $d \mid a$ and $a \mid b \Rightarrow a = dm$ and $b = an \Rightarrow b = (dm)n = d(mn) = dk$ where $k = mn \Rightarrow d \mid b$.

5. a. $836 = 33(25) + 11$
   b. $780 = -16(-48) + 12$

Exercises 6.2

1. b. $c = (a,b) \Rightarrow c \mid a$ and $c \mid b \Rightarrow c \mid (|a|)$ and $c \mid (|b|)$. Let $x \mid (|a|)$ and $x \mid (|b|)$ where $x \in \mathbf{Z}$.
   If $a \geq 0$, then $|a| = a$ and $x \mid a$. If $a < 0$, then $|a| = -a$, $x \mid (-a)$ and hence $x \mid a$.
   In like manner, $x \mid b$. Since $c = (a,b)$, $x \mid c$. Therefore $c = (|a|, |b|)$.

   c. Let $a \in \mathbf{Z}$, $a \neq 0$. Now $a \mid a$ and $a \mid 0$. Hence $(|a|) \mid a$ and $(|a|) \mid 0$. Let $x \mid a$ and $x \mid 0$.
   Then $x \mid (|a|)$. Therefore, $|a| = (a,0)$.

3. a. Let $a \neq 0$, $b \neq 0$. Then a and b have a gcd, c. Let $m = \frac{|ab|}{c}$. Then m is the lcm of a and b.

   b. $m = \langle a,b\rangle \Rightarrow a \neq 0$ and $b \neq 0 \Rightarrow$ a and b have a gcd, c. c unique $\Rightarrow \frac{|ab|}{c} = \langle a,b\rangle = m$ is unique.

   c. $m = \langle a,b\rangle = \frac{|ab|}{(a,b)} = \frac{|(|a|)(|b|)|}{(|a|,|b|)} = \langle |a|,|b|\rangle$

5. a. $(210,-240) = (210,240)$
   $240 = 210 \cdot 1 + 30$
   $210 = 30 \cdot 7 + 0$
   Therefore, $30 = (210,240) = (210,-240)$

   b. $30 = 240 - 210 \cdot 1$
   $= 210\,(-1) + 240\,(1)$
   $= 210\,(-1) + (-240)(-1)$

c. $<210,-240> \frac{|(210)(-240)|}{(210,-240)} = \frac{50400}{30} = 1680$

7. Let (m,n), (r,s) $\in$ **N x N** such that (m,n) $\neq$ (r,s). Since (m,n) $\neq$ (r,s), m $\neq$ r or n $\neq$ s.
We may assume m $\neq$ r. Then $a^m \neq a^r$ since f: **N** $\to$ **N** where $f(k) = a^k$ is 1–1.
We wish to show $a^m b^n \neq a^r b^s$. Assume $a^m b^n = a^r b^s$.

   If n = s, then $b^n = b^s$. Hence, if $a^m b^n = a^r b^s$, then $a^m b^n = a^r b^n$ or $a^m = a^r$. Contradiction.
   If n $\neq$ s, we may assume n < s. Hence, s – n > 0, $a^m = a^r b^{s-n} = b(a^r b^{s-n-1})$ and b | a^m.
   Since (a,b) = 1, b | a. But b $\geq$ 2 and thus (a,b) $\neq$ 1. Contradiction.

In either case a contradiction resulted from assuming $a^m b^n = a^r b^s$.
Therefore, $a^m b^n \neq a^r b^s$ or h((m,n)) $\neq$ h((r,s)); that is, h is 1–1.

Exercises 6.3

1. a. Let p be prime. k | (-p) $\Rightarrow$ k | p $\Rightarrow$ k = $\pm$ 1 or k = $\pm$p $\Rightarrow$ -p is prime.

3. a. If 201 is a composite integer, then there exists a prime integer p such that p | 201 and p $\leq \sqrt{201}$.
Hence, $p^2 \leq 201$. Now $2^2 = 4$, $3^2 = 9$, $5^2 = 25$, $7^2 = 49$, $11^2 = 121$, $13^2 = 169$ and $17^2 = 289$.
Hence, p must be 2, 3, 5, 7, 11 or 13. If p = 3, 3 | 201 and thus 201 is a composite integer.

   b. If 293 is a composite integer, then there exists a prime integer p such that p | 293 and p $\leq \sqrt{293}$.
Hence, $p^2 \leq 293$. Now $2^2 = 4$, $3^2 = 9$, $5^2 = 25$, $7^2 = 49$, $11^2 = 121$, $13^2 = 169$, $17^2 = 289$ and $19^2 = 361$. Hence, p must be 2, 3, 5, 7, 11,13 or 17. But $p \nmid 293$ where p = 2, 3, 5, 7, 11, 13, 17.
Thus, 293 is a prime integer.

5. a. $112 = 2^4 \bullet 7$ and $264 = 2^3 \bullet 3 \bullet 11$

   b. $\{2,7\} \cup \{2,3,11\} = \{2,3,7,11\}$
$112 = 2^4 \bullet 3^0 \bullet 7^1 \bullet 11^0$ and $264 = 2^3 \bullet 3^1 \bullet 7^0 \bullet 11^1$
$c = (112,264) = 2^3 \bullet 3^0 \bullet 7^0 \bullet 11^0 = 2^3 = 8$
$m = <112,264> = 2^4 \bullet 3^1 \bullet 7^1 \bullet 11^1 = 3696$

   c. Same results.

Exercises 6.4

1. Let $m = 2$ and $a, b \in \mathbf{Z}$.

   a. $a, b \in \mathbf{E}$ or $a, b \in \mathbf{0} \Rightarrow a, -b \in \mathbf{E}$ or $a, -b \in \mathbf{0} \Rightarrow a + (-b) = a - b \in \mathbf{E} \Rightarrow$ $a - b = 2k \Rightarrow 2 \mid (a - b) \Rightarrow a \equiv b \pmod{2}$.

   b. $a \in \mathbf{E}$ and $b \in \mathbf{0} \Rightarrow a \in \mathbf{E}$ and $-b \in \mathbf{0} \Rightarrow a + (-b) = a - b \in \mathbf{O} \Rightarrow a - b \neq 2k \Rightarrow$ $2 \nmid (a - b) \Rightarrow a \not\equiv b \pmod{2}$.

3. a. $a \equiv c \pmod{m}$ and $b \equiv d \pmod{m} \Rightarrow m \mid (a - c)$ and $m \mid (b - d) \Rightarrow$ $m \mid [(a - c) + (b - d)] \Rightarrow m \mid [(a + b) - (c + d)] \Rightarrow (a + b) \equiv (c + d) \pmod{m}$.

5. Let $ab \equiv ac \pmod{m}$ and $(a,m) = 1$. $ab \equiv ac \pmod{m} \Rightarrow m \mid (ab - ac) \Rightarrow m \mid a(b - c) \Rightarrow m \mid (b - c)$ since $(a,m) = 1$. $\therefore\ b \equiv c \pmod{m}$.

7. Let $\overline{a}, \overline{b}, \overline{c} \in \mathbf{Z}_m$

   b.
$$
\begin{aligned}
(\overline{a} + \overline{b}) + \overline{c} &= \overline{a + b} + \overline{c} \\
&= \overline{(a + b) + c} \\
&= \overline{a + (b + c)} \\
&= \overline{a} + \overline{b + c} \\
&= \overline{a} + (\overline{b} + \overline{c})
\end{aligned}
$$

c. $\overline{a} + \overline{0} = \overline{a + 0} = \overline{a}$; and, since + is commutative, $\overline{0}$ is the identity of $\mathbf{Z}_m$ relative to +.

e. $\overline{a} \bullet \overline{b} = \overline{a \bullet b} = \overline{b \bullet a} = \overline{b} \bullet \overline{a}$

h.
$$
\begin{aligned}
\overline{a} \bullet (\overline{b} + \overline{c}) &= \overline{a} \bullet \overline{b + c} \\
&= \overline{a \bullet (b + c)} \\
&= \overline{a \bullet b + a \bullet c} \\
&= \overline{a \bullet b} + \overline{a \bullet c} \\
&= \overline{a} \bullet \overline{b} + \overline{a} \bullet \overline{c}
\end{aligned}
$$

Since • is commutative, • is distributive over +.

Chapter 7
Exercises 7.1

1. Let f: **Z** → H where f(k) = 3k.
   f is 1–1:
   Let m, n ∈ **Z** such that f(m) = f(n).
   Then 3m = 3n or m = n.
   f is onto:
   Let y ∈ H. Then y = 3m where m ∈ **Z**.
   Now f(m) = 3m = y.
   Hence, f is 1–1 and onto.
   Therefore, **Z** ~ H.

3. Let A ~ B, C ~ D, A ∩ C = ∅ and B ∩ D = ∅. There exists functions f: A → B and g: B → D which are 1–1 correspondences.
   Let f ∪ g: A ∪ C → B ∪ D where

$$(f \cup g)(x) = \begin{cases} f(x) \text{ when } x \in A \\ g(x) \text{ when } x \in C \end{cases}$$

   f ∪ g is 1–1:
   Let r, s ∈ A ∪ C such that (f ∪ g)(r) = (f ∪ g)(s).
   Now (f ∪ g)(r), (f ∪ g)(s) ∈ B ∪ D; and, since B ∩ D = ∅, (f ∪ g)(r), (f ∪ g)(s) ∈ B or (f ∪ g)(r), (f ∪ g)(s) ∈ D but not both. Thus, (f ∪ g)(r) = f(r) and (f ∪ g)(s) = f(s) or (f ∪ g)(r) = g(r) and (f ∪ g)(s) = g(s). Hence, f(r) = f(s) or g(r) = g(s). In either case r = s since f and g are 1–1. Therefore, f ∪ g is 1–1.

   f ∪ g is onto:
   Let y ∈ B ∪ D. Since B ∩ D = ∅, y ∈ B or y ∈ D but not both.
   If y ∈ B, ∃ a ∈ A such that f(a) = y. If y ∈ D, ∃ c ∈ C such that g(c) = y.
   Hence, a ∈ A ∪ C and (f ∪ g)(a) = f(a) = y or c ∈ A ∪ C and (f ∪ g)(c) = g(c) = y.
   Therefore, f ∪ g is onto.

   Hence, f ∪ g is 1–1 and onto.
   Therefore, A ∪ C ~ B ∪ D.

Exercises 7.2

1. Since $A \sim B$, there exists a function $f: A \longrightarrow B$ which is a 1–1 correspondence. Hence, $A \neq \emptyset$. If A if finite, then $\mathbf{N}_k \sim A$. Therefore, we have $\mathbf{N}_k \sim A$ and $A \sim B$. Thus $\mathbf{N}_k \sim B$ and B is finite.

3. For $n \geq 2$, let p(n):

   If $A_1, A_2, \ldots, A_n$ are finite sets, then $A_1 x A_2 x \ldots x A_n$ is a finite set.

   p(2): Let A_1, A_2 be finite sets. Then $A_1 \text{ x } A_2$ is a finite set by Theorem 7.2.5. Thus p(2) is true.

   p(k): Assume p(k) is true; i.e., if $A_1, A_2, \ldots, A_k$ are finite sets, then $A_1 \text{ x } A_2 x \ldots x A_k$ is a finite set.

   p(k + 1): Show p(k + 1) is true. Let $A_1, A_2, \ldots, A_k, A_{k+1}$ be finite sets.
   Now $A_1 x A_2 x \ldots x A_k x A_{k+1} \sim (A_1 x A_2 x \ldots x A_k) \text{ x } A_{k+1}$.
   Since $A_1 x A_2 x \ldots x A_k$ is finite and A_{k+1} is finite, $(A_1 x A_2 x \ldots x A_k) \text{ x } A_{k+1}$ is finite (Theorem 7.2.5). By Problem (1) above, $A_1 x A_2 x \ldots x A_k x A_{k+1}$ is finite.
   Thus p(k + 1) is true.

   Therefore, $(\forall n, n \geq 2)(p(n))$ is true.

Exercises 7.3

1. Let A be an infinite set. Assume B is a finite set.
   a. If $A \sim B$, then by Exercises 7.2, Problem (1), A is finite. Contradiction. Hence B is infinite.
   b. If $A \subsetneq B$, then by Theorem 7.2.4, A is finite. Contradiction. Hence, B is infinite.

3. Let $A \sim B$. If A is denumerable, then $\mathbf{N} \sim A$. Therefore, we have $\mathbf{N} \sim A$ and $A \sim B$. Thus $\mathbf{N} \sim B$ and B is denumerable.

5. Let A be an infinite set. Show A has a denumerable subset B.
   If A is denumerable, let B = A.
   Suppose A is not denumerable.
   Pick any element in A and denote it by a_1.
   Pick any element in $A - \{a_1\}$ and denote it by a_2.
   Pick any element in $A - \{a_1, a_2\}$ and denote it by a_3.
   ⋮ ⋮
   Pick any element in $A - \{a_1, \ldots, a_{n-1}\}$ and denote it by a_n.

Since A and $A - \{a_1, \ldots, a_{n-1}\}$ are infinite sets (see Problem (4)), we can continue the process. Let $B = \{a_1, a_2, \ldots, a_n, \ldots\}$. Then B is a denumerable set; and, since A is not denumerable, $A - B \neq \emptyset$ and hence $B \subset A$.

7. Let A be denumerable and $B = \{b_1, \ldots, b_n\}$. For each $b_i \in B$, $A \times \{b_i\} \sim A$.

Hence, $A \times \{b_i\}$ is denumerable. By Theorem 7.3.6, $\bigcup_{i=1}^{n} (A \times \{b_i\}) = A \times (\bigcup_{i=1}^{n} \{b_i\}) = A \times B$

is a denumerable set.

Exercises 7.4

1. For $n \geq 2$, let p(n): If $A_1, A_2, \ldots, A_n$ are countable sets, then $\bigcup_{i=1}^{n} A_i$ is a countable set.

p(2): Let A_1, A_2 be countable sets. By Theorem 7.4.2, $A_1 \cup A_2$ is a countable set. Hence, p(2) is true.

p(k): Assume p(k) is true; i.e., if $A_1, A_2, \ldots, A_k$ are countable sets, then $\bigcup_{i=1}^{k} A_i$ is a countable set.

p(k + 1): Show p(k + 1) is true. Let $A_1, A_2, \ldots, A_k, A_{k+1}$ be countable sets.
Now $\bigcup_{i=1}^{k+1} A_i = (\bigcup_{i=1}^{k} A_i) \cup A_{k+1}$. Since $\bigcup_{i=1}^{k} A_i$ and A_{k+1} are countable sets, by Theorem 7.4.2, $(\bigcup_{i=1}^{k} A_i) \cup A_{k+1}$ is a countable set.
Hence, $\bigcup_{i=1}^{k+1} A_i$ is countable and p(k + 1) is true.

Therefore, $(\forall n, n \geq 2)$ (p(n)) is true.

Chapter 8
Exercises 8.1

1. $\mathcal{R}$ is a partial order on **N**.

   Let $a \in \mathbf{N}$. Since $a = a \cdot 1$, $a \mid a$. Hence, $a \mathcal{R} a$ and $\mathcal{R}$ is reflexive.

   Let $a, b \in \mathbf{N}$ such that $a \mathcal{R} b$ and $b \mathcal{R} a$. Then $a \mid b$ and $b \mid a$ or $b = ap$ and $a = bq$ for $p, q \in \mathbf{N}$. Therefore, $a = bq = (ap)q = a(pq)$ and $1 = pq$. Hence, $p = q = 1$ and $a = b$. Thus, $\mathcal{R}$ is anti-symmetric.

   Let $a, b, c \in \mathbf{N}$ such that $a \mathcal{R} b$ and $b \mathcal{R} c$. Then $a \mid b$ and $b \mid c$; or, $b = ap$ and $c = bq$ for $p, q \in \mathbf{N}$. Thus, $c = bq = (ap)q = a(pq)$. Therefore, $a \mid c$ and $a \mathcal{R} c$. Thus, $\mathcal{R}$ is transitive.

   $\mathcal{R}$ is not a total order of **N** since $2, 3 \in \mathbf{N}$ but $2 \not\mathcal{R}\, 3$ and $3 \not\mathcal{R}\, 2$.

3. $\mathcal{R}$ is a partial order on P(X).

   Let $A \in P(X)$. Since $A \subseteq A$, $A \mathcal{R} A$ and hence $\mathcal{R}$ is reflexive.

   Let $A, B \in P(X)$ such that $A \mathcal{R} B$ and $B \mathcal{R} A$. Then $A \subseteq B$ and $B \subseteq A$. Thus, $A = B$ and $\mathcal{R}$ is anti-symmetric.

   Let $A, B, C \in P(X)$ such that $A \mathcal{R} B$ and $B \mathcal{R} C$. Then $A \subseteq B$ and $B \subseteq C$. Thus, $A \subseteq C$ and $A \mathcal{R} C$. Therefore $\mathcal{R}$ is transitive.

   $\mathcal{R}$ is not a total order on P(X) since $A, A' \in P(X)$, but $A \not\mathcal{R}\, A'$ and $A' \not\mathcal{R}\, A$.

5. a. $a \leq b \Rightarrow b - a$ is non-negative. Now $b - a = (b + c) - (a + c)$. Hence, $(b + c) - (a + c)$ is non-negative and thus $a + c \leq b + c$.

   b. $a \leq b \Rightarrow b - a$ is non-negative. For c positive, $(b - a)c = bc - ac$ is non-negative and thus $ac \leq bc$.

   c. $a \leq b \Rightarrow b - a$ is non-negative. For c negative, $(b - a)c = bc - ac$ is non-positive. Hence, $-(bc - ac) = ac - bc$ is non-negative and thus $bc \leq ac$ or $ac \geq bc$.

Exercises 8.2

1. Let l = glb A and l' = glb A. Now l' is a lower bound of A; and, since l = glb A, $l \leq$ 1′. Also l is lower bound if A; and, since l' = glb A, $l \leq l'$. Therefore, since $\leq$ is anti-symmetric, $l = l'$. Thus, if l = glb A, then l is unique.

3. a. Let u = lub A and $\delta > 0$. Now $u - \delta < u$. If there does not exist $a \in A$ such that $u - \delta < a < u$, then $a \leq u - \delta$, $\forall a \in A$. Hence, $u - \delta$ is an upper bound of A; and, since $u - \delta < u$, $u \neq$ lub A. Contradiction.

   b. Let l = glb A and $\delta > 0$. Now $l < l + \delta$. If there does not exist $a \in A$ such that $l < a < l + \delta$, then $l + \delta \leq a$, $\forall a \in A$. Hence, $l + \delta$ is an lower bound of A; and, since $l < l + \delta$, 1 $\neq$ glb A. Contradiction.

5. Let $A \subseteq \mathbf{R}$, $A \neq \emptyset$. Let l be a lower bound of A. Then $l \leq a$, $\forall a \in A$. Thus, $-l \geq -a$ and $-l$ is an upper bound of B where $B = \{-a \mid a \in A\}$. Hence, by Axiom of Completeness, B has a lub, say u. By problem (4), $-u$ = glb A.

Exercises 8.3

1. Let $\delta > 0$. Then 1 and δ are positive real numbers. By the Archimedean Property there exists $n \in \mathbf{N}$ such that $1 < n\delta$ or $\frac{1}{n} < \delta$.

3. Let $a, b \in \mathbf{Q}$ such that $a < b$. Then there exists an irrational number r such that $a < r < b$. Let $A = \{x \in \mathbf{Q} \mid x < r\}$. $A \subseteq \mathbf{Q}$; and, since $a \in A$, $A \neq \emptyset$. Now A has an upper bound in $\mathbf{Q}$, i.e., b. However, A does not have a least upper bound in $\mathbf{Q}$. Assume there exists $u \in \mathbf{Q}$ such that u = lub A. Since $u \in \mathbf{Q}$, $u \neq$ r. Hence, either $u < r$ or $u > r$. In either case, there exists $c \in \mathbf{Q}$ such that $u < c < r$ or $r < c < u$. Thus $u \neq$ lub A. Contradiction. Therefore, no such $u \in \mathbf{Q}$ exists.

5. $I_i \supseteq I_{i+1}$:

   $x \in I_{i+1} = [i+1, \infty) \Rightarrow x \geq i + 1 > i \Rightarrow x \in I_i = [i, \infty)$.

   $\bigcap_{i \in \mathbf{N}} I_i = \emptyset$:

   $\bigcap_{i \in \mathbf{N}} I_i \neq \emptyset \Rightarrow \exists\, x \in \bigcap_{i \in \mathbf{N}} I_i \Rightarrow x \in I_i = [i, \infty)\ \forall\, i \in \mathbf{N}$.

   Now x, $1 \in \mathbf{R}^+$. Hence $\exists\, n \in \mathbf{N}$ such that $x < n \cdot 1 = n$. Therefore, $x \neq I_n = [n, \infty)$. Contradiction.

   Thus $\bigcap_{i \in \mathbf{N}} I_i = \emptyset$.

Index

E

F

G

I

S

T

U

V

W